S0-DVV-683

A PRIMER IN ECOTHEOLOGY

CASCADE COMPANIONS

The Christian theological tradition provides an embarrassment of riches: from Scripture to modern scholarship, we are blessed with a vast and complex theological inheritance. And yet this feast of traditional riches is too frequently inaccessible to the general reader.

The Cascade Companions series addresses the challenge by publishing books that combine academic rigor with broad appeal and readability. They aim to introduce nonspecialist readers to that vital storehouse of authors, documents, themes, histories, arguments, and movements that comprise this heritage with brief yet compelling volumes.

SELECTED TITLES IN THIS SERIES:

Reading Paul by Michael J. Gorman
Theology and Culture by D. Stephen Long
Creation and Evolution by Tatha Wiley
Theological Interpretation of Scripture by Stephen Fowl
Reading Bonhoeffer by Geffrey B. Kelly
Justpeace Ethics by Jarem Sawatsky
Feminism and Christianity by Caryn D. Griswold
Angels, Worms, and Bogeys by Michelle A. Clifton-Soderstrom
Christianity and Politics by C. C. Pecknold
A Way to Scholasticism by Peter S. Dillard
Theological Theodicy by Daniel Castelo
The Letter to the Hebrews in Social-Scientific Perspective by David A. deSilva
Basil of Caesarea by Andrew Radde-Galwitz
A Guide to St. Symeon the New Theologian by Hannah Hunt
Reading John by Christopher W. Skinner
Forgiveness by Anthony Bash
Jacob Arminius by Rustin Brian
Jeremiah: Prophet Like Moses by Jack Lundbom
John Calvin by Donald McKim
Rudolf Bultmann: A Companion to His Theology by David Congdon
The U.S. Immigration Crisis: Toward and Ethics of Place by Miguel A. De La Torre
Theologia Crucis: A Companion to the Theology of the Cross by Robert Cady Saler
Virtue: An Introduction to Theory and Practice by Olli-Pekka Vainio
Approaching Job by Andrew Zack Lewis

A PRIMER IN ECOTHEOLOGY

Theology for a Fragile Earth

CELIA DEANE-DRUMMOND

CASCADE *Books* • Eugene, Oregon

A PRIMER IN ECOTHEOLOGY
Theology for a Fragile Earth

Cascade Companions 40

Cascade Books
An Imprint of Wipf and Stock Publishers
199 W. 8th Ave., Suite 3
Eugene, OR 97401

www.wipfandstock.com

PAPERBACK ISBN: 978-1-4982-3699-7
HARDCOVER ISBN: 978-1-4982-3701-7
EBOOK ISBN: 978-1-4982-3700-0

Cataloguing-in-Publication data:

Names: Deane-Drummond, Celia, author.

Title: A primer in ecotheology : theology for a fragile earth / Celia Deane-Drummond.

Description: Eugene, OR: Cascade Books, 2017 | Cascade Companions 37 | Includes bibliographical references.

Identifiers: ISBN 978-1-4982-3699-7 (paperback) | ISBN 978-1-4982-3701-7 (hardcover) | ISBN 978-1-4982-3700-0 (ebook)

Subjects: LCSH: Human ecology—Religious aspects—Christianity | Environmental protection—Moral and ethical aspects.

Classification: BF353.N37 D 2017 (print) | BF353 (ebook).

Manufactured in the U.S.A. 07/23/19

For Dara

31st October 2013—22nd June 2017.

CONTENTS

PREFACE

THIS BOOK STEMS FROM my own experience teaching classes in ecotheology both at the University of Chester (UK) since 1994 and the University of Notre Dame (USA) since 2011. The first book I ever published in 1996 was sponsored by the World Wide Fund for Nature under the auspices of my work as a part-time consultant with the International Consultancy of Religion, Education and Culture (ICOREC) in Manchester, UK, and is titled *A Handbook in Theology and Ecology.* That book, like this one, is intended to be a user-friendly primer in ecotheology. Since I wrote that book, scientific aspects of climate change and biodiversity loss have become even more serious. There are other concerns that are new developments in the field.

The *Handbook,* like this book, dealt with core areas relevant to all Christian readers, namely, how to interpret the Bible from an ecologically aware context. But since that book was written, the interpretation of Scripture from an ecological perspective has developed and enlarged into a much bigger field of biblical ecological hermeneutics. In the *Handbook* I dealt with Celtic Christianity as a way of resonating with the historical aspects of the faith most relevant to my contextual location in the UK. In this book I address agrarianism that has been both particularly significant in the USA, and also habitually practiced in Europe as well. Further, while I do engage tangentially with political issues, the contextual limitations and practical restraints for a book of this size are obvious. I have not, for example, included much on ecclesial or liturgical aspects that are also an important dimension for ecotheology. I have, however, deliberately represented scholarship from different Christian denominational backgrounds.

Ecotheology is aware of itself as situated theology, and that means recognizing that while as a writer I include diverse global perspectives, my own views are necessarily colored by my background as a Caucasian woman scientist who has trained in a particular Western tradition in the natural sciences, has lived in privileged places in the world, and now teaches at a private university in the United States that has an even bigger endowment than some of the small nation states that are impacted by climate change. It is hard to avoid these paradoxes. In the *Handbook*, I devoted specific chapters to the Gaia hypothesis, ecological liturgy, and ecofeminist thought. In this book I have devoted a specific chapter to consider the significance of Pope Francis's 2015 encyclical, *Laudato Si'*. I have named him as *icon of the Anthropocene* not least because he represents a different way of thinking about what it means to be human, with relevance that goes far beyond the confines of the Roman Catholic Church.

I have paid more attention in this book to the specific systematic elements in ecotheology by giving a more in depth discussion to two crucial areas, namely, Christology and anthropology. Discussion of a doctrine of creation is of course also important, and rather more obvious, and elements of those insights are woven into the chapter on *Ecological Biblical Hermeneutics* and other places. Christology also raises critical issues about how to deal with suffering, explicitly, ecological suffering and future redemption. The systematic chapters are intended to push students to think through standard systematic treatments of Christian doctrine in the light of ecological concerns.

I am conscious that in writing this book there is so much more that I could have done; this book is really a sampling of ideas rather than anything more comprehensive. For example, I have not delved into all the Christian historical texts that are relevant to this field, or covered all the different areas of environmental ethics and background philosophical debates. I hope that my academic colleagues who, like me, have labored in this field for the best part of a quarter century or longer, will forgive me if some of their most treasured ideas or even books or articles do not appear here. My intention in this book is to give students and other readers just a taste for what this is about, a preview of what can be developed in further research, rather than try and cover everything that might be relevant. My own experience of teaching is that the level of ignorance is very high, and there are hard choices to make when deciding where to start. But my hope is to combine both depth in some areas and breadth in others, so that readers can see both the intellectual challenges and also the scope of this horizon.

I have also added a glossary of key terms in ecotheology in order to help those who are using this book gain a better understanding of what these terms mean, and to save space on definitions every time a term appears. One of the movements in ecotheology in recent years has been a growing alliance with eco-criticism that has developed in English literature. The relationship with those working in religious studies and environmental questions is also rather more amenable than perhaps in the past, mostly because, given the size and scope of the problems to be addressed, dialogue and collaboration is vital. This is the message of Pope Francis as well, that committed Christians cannot afford *not* to take ecology seriously, and that we need to work with all those of different religious faiths, or no faith, in order to address what is arguably the most pressing problem of our generation.

This book is quite deliberately not a book on religious environmentalism in a very general sense, though I do point to some further resources in that field in terms of religiously inspired environmental practice in the final appendix. My own view is that Christian believers need to understand more about their own traditions and their relevance to this field prior to engaging in sensitive dialogue with those from other faiths or none that share their concerns for the planet, for those living in poverty already impacted by climate change, and for future generations.

ACKNOWLEDGMENTS

I AM GRATEFUL TO numerous colleagues in the field of ecotheology, as well as all the public audiences at various lectures I have given and students who have helped shape the way this book has developed. Special thanks go to the small class of undergraduate and graduate students at the University of Notre Dame, who took my Fall 2016 course titled *Theology for a Fragile Earth*. Allison Zimmer, Patrick Yerkes, Thomas Wheeler, Kyle Planck, Julia Henkel, and Rachel Francis were undergraduate students and Kyle Nicholas, Sr. Juliet Namiiro, Marie Claire Klassen, and Sebastian Ekberg the graduates. Their questions, insights, curiosity, and desire to become engaged in practical action was a source of inspiration for me in writing this book, and each and every one of them contributed something to the way this book ended up. I am also particularly grateful to my teaching and research assistant for that semester, Michelle Marvin, who has also helped with copy editing and compiled the index. My colleague at Notre Dame, Fr. Terrence Ehrman, C.S.C, trained as a stream ecologist, guided our field trip to Warren Woods and reminded me of the need to experience the living reality of natural world as a way of inspiring the intention to care. The questions for discussion in this book, at least on occasions, try and encourage that practice. I am also grateful to Christian Amondson of Wipf and Stock, who first invited me to write this book in 2015, and to K. C. Hanson and other members of the Wipf and Stock editorial team. Special thanks too for support by my family, my husband Henry, and my two daughters Sara and Mair, but not least our black English Labrador dog, Dara. Dara means *full of compassion* in Hebrew, and that, it seems to me, is one of the driving motivations that spur many of us into considerations of serious threats to people and planet. Our companion animals also remind us that human societies are bound up with the lives of other creatures for whom we have special and unique responsibilities.

Tragically, as this book went to press, Dara died dramatically on the day of our arrival in the United Kingdom. The paradox of her ending, cut short in her prime, is a constant reminder of the fragility of life. Yet, her humility, devotion, and patience in her suffering witnesses to her inner strength and ability to evoke deep compassion in us. May she R.I.P.

1

INTRODUCTION TO ECOTHEOLOGY: A MAP

This chapter lays out the core methodological issues in approaching ecotheology, including key aspects of climate change and ecological science, the experience of ecological devastation at a local and global level, and a broad framework for a theological ecological ethics.

WHAT HAS ECOLOGY GOT to do with theology? I used to be asked that question as a younger scholar about thirty years ago when I first started working at the boundary between theology and ecological issues. That question still comes up rather frequently today. Devastating climate change impacts the poorest of the poor living in some of the most deprived areas of the world and there are disproportionate environmental harms in deprived regions of most mega-cities across the world. The result, environmental injustice, means that theology is definitely concerned with such matters. There are also more theologians now interested in creation and scientific matters; perhaps the cultural idealization of science has given way to a greater awareness that science is not value free, and therefore the territory that it covers is something that theologians can become concerned about.

But there remains an uncomfortable gap, even among those that call themselves ecotheologians. Although it might seem obvious that those who care about environmental degradation, loss of biodiversity, and climate change would refer to climate science, many of those working in ecotheology do not dwell on the scientific aspects. Why? Is there nervousness about competency? Or, more likely, is it that the particular contribution from the humanities is different from that of science, so any examination of the credibility of that science seems to be out of place?

Another core characteristic of ecotheology, which will be obvious when reading this book, is that it covers a range of different theological standpoints and positions. For example, an ecotheologian might draw on very traditional theological resources, such as Francis of Assisi or Maximus the Confessor, or she might align herself with feminist thinkers and scholars, or even radical political activists. As the theological boundaries are pushed wider open, so are the range of critical analyses, from specific individual virtues that can be explored, such as frugality, practical wisdom, and temperance through to a focus on flaws in political and social structures that lead to environmental and other harms. The oscillation between the local and the global gives ecotheology a particular dynamic that is different from other contextual theologies. But even this aspect is contested among ecotheologians. Global problems such as climate change seem to be so difficult to solve that many have argued that it is time to target local, regional, and specifically agricultural methods.

I am one of a handful of ecotheologians who believe that in order to do ecotheology in a responsible way, we are beholden to try and take at least some account of what is the most common consensus in the scientific community on relevant topics, even if that is going to change over time. Part of the complexity of this issue in a US context is that climate change, for example, has become a highly charged and highly contested political issue. Democrats generally support arguments for humanly induced climate change, while the majority of climate change deniers are Republicans. The majority of climate scientists watch in bemusement at the way this serious scientific topic has been taken over as a political football—at least in North America and Australia. Such a bifurcation along party political lines is not characteristic of other regions of the world, including most of Europe, Africa, and Asia.

Scientists will agree, of course, that in so far as it is possible to discern, particular values will be embedded in the way science is done. But what they resist, and vociferously so, is that scientific data is somehow invented or constructed in order to serve political or ideological ends. Where science does that, it ceases to be science and becomes something else, as the history of eugenics, for example, testifies. Social scientists may examine the social background and biases that are built into scientific practice, but to claim that scientific data is simply emerging out of those presuppositions, rather than telling us something about the world as such, would be to undercut the whole driving purpose and success of science. I am therefore deliberately going to start this book with a brief overview of aspects of that science in order to provide a scientific background to ecotheology, and follow this by charting out historically the way ecotheology has developed over the last thirty years.

CLIMATE CHANGE

Calculating the warming of the planet due to the trapping of heat by carbon dioxide is routine for astronomy students. The planet Venus, which has a thick carbon dioxide atmosphere, reaches a surface temperature of around 4700 centigrade. Other greenhouse gases such as methane further contribute to warming on planet Earth. Climate models are complicated by so many different factors necessary for consideration. Local weather effects are harder to attribute, but overall trends can be traced to specific climate changes. The scientific consensus is that human beings are changing the climate both in the short and long term, often in unpredictable ways, called a climate *weirding*. Its short term impacts lead to more frequent storms, floods, droughts, and fires; its long term effects result in rising ocean levels, deep ocean heating, and ocean acidification. The evidence for global warming is solid and has been steadily accumulating ever since the mid-nineteenth century. The uncertainty in the models used by climate scientists has sometimes been misinterpreted by policy makers as uncertainty over the impacts of human activity on climate change, but given the number of variables that are used in climate science, it is almost impossible to provide absolutely accurate predictions. The International Panel on Climate Change (IPCC) has published estimates that are in the right ballpark in a predictive sense for temperature rise but underestimate global sea level rise.[1]

BOX 1: GLOBAL WARMING AND ITS IMPACTS

The best evidence for global warming comes from analysis of ice cores drilled from deep within Antarctic glaciers, which are also rapidly melting. These ice cores include small bubbles of air that contain the trapped carbon dioxide of that era, hence allowing a direct tracking of the levels of carbon dioxide back in time. The ratios of isotopes in the water in these samples also allow scientists to infer global average past temperatures. These methods show that carbon dioxide and temperature levels have varied in direct relationship with each other for the last nine ice ages, and the amount of carbon dioxide in the atmosphere is higher now than it has been for the last 800,000 years. There are natural variations in the tilt of the planet's rotation axis that contributes to driving the Earth's climate system into periodic ice ages, roughly every 100,000 years. Natural ice ages that are mapped in these ice core studies always show temperature changes *prior* to changes in carbon dioxide levels, since the shifts in the earth's orbital cycles decrease solar heating which triggers a change in ocean biology, so more carbon dioxide is absorbed in lower temperatures, leading to subsequent lower carbon dioxide levels. The biological responses to these changes are among some of the most catastrophic to consider and are often not raised. For example, vegetation on land and phytoplankton in the oceans in higher temperatures reduce their net uptake of carbon dioxide as the rates of respiration (which release carbon into the atmosphere) outpaces the rate of photosynthesis (which traps carbon). The overall decrease in oxygen could have catastrophic impacts on the ability of planet Earth to support life forms that rely on oxygen, including humanity.

Sakimoto, "Understanding the Science," *in press*.

1. Intergovernmental Panel on Climate Change, "Summary." See also U.S. Environmental Protection Agency, "Climate Change."

BIODIVERSITY LOSS

One of the impacts of climate change that has attracted significant attention from ecotheologians, in addition to a range of other impacts on ecological functioning,[2] is the loss of different species. Life on planet Earth has existed from the first bacterial stromatolite colonies that dated from 3.7 billion years ago in the Earth's 4.5 billion-year history. The background rate of extinction through the process of evolution can be calculated and compared with present day rates. The ignorance of different biological forms is perhaps even greater than that of climate change mentioned above: in a single handful of soil there are literally thousands of species present.[3] In addition only about 1.5 million of the world's estimated 8.7 million species have been identified.[4] In spite of these limitations the best estimates for the present rate of extinctions are horrifying; these rates are several orders of magnitude higher than in the past, leading some scholars to claim that we are living through the sixth great extinction event. This is comparable to other historical catastrophic extinction events, each of which required about ten million years of recovery time. The difference in this case is that extinctions are the result of human activity, either indirectly through climate change and loss of habitat, or directly through destruction for particular human uses. The International Union for the Conservation of Nature (IUCN) tracks the loss of species. Some of the latest results for those species on the "red" list that are threatened with imminent extinction were released at the World Conservation Congress held

2. For an accessible overview of ecological issues relevant to theological discussion see Ehrman, "Ecology."

3. Wilson, "A Cubic Foot."

4. Mora et al., "How Many Species."

in Hawaii in 2016. There are also those species that are extinct in the wild (for example the species may be kept in captivity in zoos), or they may be endangered, which is the third level of threat, or vulnerable, a fourth level. For example, the world-wide giraffe population has plummeted by 40% over the last thirty years and has now been listed as a vulnerable species. 11% of the 700 newly identified bird species are now threatened with extinction. At the 13th Conference of Parties to the Convention on Biological Diversity (CBD COP 13) in December 2016, the Director General of IUCN Inger Andersen remarked that "many species are slipping away before we can even describe them." The threat of extinction for wild strains of barley, mango, and oats has severe consequences for food security, since they provide a basis for genetic variation and resistance to disease, drought, and salinity through breeding and other crop programs.[5]

BOX 2: RESILIENCE OF ECOSYSTEMS

The philosophy of ecology has shifted in the last hundred years. The tendency to describe ecological systems in terms of stability, with humanity as outside observers to that system, has been replaced with the language of flux, where humans are included in that ecosystem.[6] Some stability is required for the ecosystem to function at all, and this is defined as the *resistance* of the system to environmental perturbation. Global climate change impacts on ecosystems and their ability to function well in a way that has not been the case in the past. *Resilience* is the term ecologists use to describe the way an ecosystem that has been disturbed is capable of returning to its previous state. The total number of species

5. International Union for Conservation of Nature (IUCN), "New Bird Species."

6. For further discussion see Deane-Drummond, *The Ethics of Nature*, 36–38.

in a given community is known as species richness, while the actual species present is known as species composition. It is this composition that impacts on the overall resilience of an ecosystem. Where there is competition for resources, if the different species in an ecosystem use these resources at different times or seasons, then the ecosystem as a whole is more productive. On the one hand, biological diversity generally increases resilience of the ecosystem as a whole to environmental changes, with some species playing a more significant role in this respect than others. On the other hand, this is complicated by the fact that specific species are more likely to be vulnerable in a highly diverse ecosystem, since the population size of any given species is lower, and its loss will not necessarily prevent continued ecosystem function. If two species share a role in a given ecosystem they are sometimes termed redundant; but in the long term this is an advantage if one of the species goes extinct. The primary importance of this work is that it shows the importance of inter-relationships in ecological understanding.

Cleland, "Biodiversity," 14.

AGRARIAN ALTERNATIVES

Agrarianism, that means in general a focus on the natural proclivity of the land and its patterns of food production, is a type of political social ecology that has attracted much attention, particularly in the US context. Agrarianism implies a sympathetic relationship with the land, the soil and its produce, and the creatures that share the land, as compared with agribusiness that understands the land simply as a resource to be exploited. Wes Jackson also adds in the argument that from a scientific point of view agrarianism makes sense, for it grows out of an understanding of how different organisms interrelate in a way that is not true for industrialized farming.[7] Further, once the slow

7. Jackson, "The Agrarian Mind."

devastation and erosion of the soil through non-agrarian methods is taken into account, the agrarian view becomes not just a cultural shift but also a "practical necessity" if soils are to continue to provide for the food that is needed for future generations.

BOX 3: THE ROOTS OF AGRARIANISM

North American writers such as Aldo Leopold, John Muir, Henry David Thoreau, and Wendell Berry have all succeeded in making agrarianism a popular standpoint. Insofar as this approach focuses on specific localized practices of farming and land management, it seems more within reach practically than attempts to reach legally binding international agreements. Agrarianism has helped foster a massive growth in farmers markets the world over, but especially in the USA. Norman Wirzba is an ecotheologian whose writing has consistently been inspired by agrarian thinking. He believes that we need to recover agrarian perspectives in order to heal the split with the land that has surfaced in urban cultures. He writes:

> Agrarian life, with its concrete and practical engagement with the forces of life and death, makes possible the intimate knowledge of and sympathy for the earth that are indispensable in the care of creation. Urban life, on the other hand, since it limits the vital connection between humanity and the earth, has the potential to thoroughly insulate us from the grace of life and health . . . The earth, on the agrarian view, is not simply a resource, but rather a source of inexhaustible life.

Wirzba, *The Paradise of God*, 72–73.

ECOTHEOLOGIES: A FIELD IN MOTION

For those writers, like myself, who have been engaged in this topic for a lifetime or more, and who bear in mind its scope from global questions about climate change to more focused studies on agrarian methods, trends within the field itself start to become clear. When ecotheology began in the late seventies and early eighties, it was more likely to be termed "green" theology rather than ecotheology, and it was more likely to have arisen, at least in Protestant circles, from a general sense of dissatisfaction with a lack of theological attention to the created order as such. Jürgen Moltmann was one of the key writers in the Protestant world who started to swing the theological pendulum towards taking the natural world seriously, with a major contribution to the discussion in *God in Creation: An Ecological Theology*.[8]

Since he began his work there has been an explosion of books, articles, and practical resources published in the field. Environmental concerns slipped into the discussion in a more subtle way in Roman Catholic thought through traditional Catholic social teaching, though key Catholic writers that parted company from that stream of thought included radical feminist scholars such as Rosemary Radford Ruether,[9] Mary Grey,[10] Anne Primavesi,[11] and more latterly Elizabeth Johnson,[12] as well as more traditional scholars such as Denis Edwards,[13] who sought to retrieve relevant material from early church fathers like Basil of Caesarea or Athanasius. Sean McDonagh, a Roman Catholic Columbian priest with a background in social science, also influenced the field by his insistence that

8. Moltmann, *God in Creation*.

9. Ruether, *Gaia and God*.

10. Grey, *Sacred Longings*.

11. Primavesi, *From Apocalypse to Genesis*.

12. Elizabeth Johnson, "Jesus and the Cosmos."

13. Edwards, *Ecology at the Heart of Faith*.

ecclesial reform was needed in order to take account of the plight of the poorest of the poor and the earth.[14] During this period the focus of discussion has started to shift, even if elements of earlier debates remain. It is useful to understand the reasons behind these shifts in order to appreciate more clearly the specific ways in which ecotheology has responded to the practical needs of the ecclesial community as well as its intention to contribute to a wider public and practical debate on environmental issues.[15]

From Anthropocentrism to Biocentrism to Theocentrism

The early years of discussion of the practical issues arising in the 1970s and 1980s were primarily about ethics and environmental responsibility considered in terms of human action and external resources: How should humans use environmental resources responsibly? Over time, this approach seemed less than adequate. Treating creatures simply as resources failed to take sufficient account of their value in and of themselves, or their *intrinsic value*. The shift to a focus on all biological organisms—or *biocentrism*—and other variants of a political position known as *deep ecology*, which sought to weave in a specific political platform for action on behalf of other creatures, was perfectly understandable in this context.[16] Other important voices in this early period included James Gustafson, who

14. McDonagh, *The Greening of the Church*. McDonagh was also a significant influence on Pope Francis's encyclical on the environment, *Laudato si'*.

15. For a discussion of these debates see Deane-Drummond and Bedford-Strohm, *Religion and Ecology*.

16. Taylor and Zimmermann, "Deep Ecology."

argued for *theocentrism* as an alternative to the divide between a human centered versus bio-centered approach.[17]

From Androcentrism to Ecofeminism

Ecofeminist writers, on the other hand, led most notably by Rosemary Radford Ruether, criticized not just too great a focus on humans—anthropocentrism—but also *androcentrism*, that is, a focus on the male half of the human race. Each contributed to doubly oppressive structures in relation to both the natural world *and* women.[18] But the association of women and nature was a double-edged sword. On the one hand, some feminists wanted to claim and celebrate their differences from men. On the other hand, the association of women and nature smacked of essentialism, a tying-in of women to particular roles that were constructed by nature rather than by nurture. Given such a dilemma, it is not surprising that ecofeminist writers are drawn to alternative ways of perceiving the natural world that avoid the anthropocentric/biocentric/theocentric triangulation. Most often this is by looking to cosmic models that seem capable of including God, humanity, and the biosphere.

One powerful model is the world as a body of God favored by Sallie McFague.[19] Another is the Gaia hypothesis pioneered by scientist and maverick James Lovelock that is used to good effect by Anne Primavesi.[20] Both

17. Gustafson, *A Sense of the Divine*.

18. Deane-Drummond, "Creation."

19. McFague, *The Body of God*.

20. Primavesi's *From Apocalypse to Genesis* has given way to books more self-consciously informed by Lovelock's Gaia hypothesis, as in Primavesi, *Sacred Gaia*.

McFague and Primavesi are prolific authors that try at least to take into account environmental science in their deliberations. They also work in a constructive manner to build new and original ways of thinking about God in the light of environmental concern. Lisa Sideris has criticized McFague for her idealistic interpretation of science, and much the same could be said of Primavesi's incorporation of Gaia.[21]

Emerging Grand Narratives of Creation

A strand that is particularly influential in the North American context could be named as the new creation story under the influence of prolific writer and Roman Catholic priest Thomas Berry. Many authors sympathetic to ecofeminism prefer this approach, including Mary Evelyn Tucker, John Grim, John Haught, Heather Eaton, and Anne Marie Dalton.[22] Here the cosmic evolution of the planet provides the framework for thinking about God, humanity, and the cosmos. For protagonists of the new creation story, also sometimes called the story of the universe, this shift seems self-evident, rather than subject to critique, but the myth making around this particular story is powerfully resonant across a range of religious traditions. But does this also portray a particular scientific understanding as the "ultimate" myth of the universe? In this respect, it is worth asking how far religious awe is being transferred to science as such, and so becoming a

21. Sideris, *Environmental Ethics*, 45–90.

22. Tucker and Swimme, *Journey of the Universe*; Grim and Tucker, *Ecology and Religion*; Haught, *The Promise of Nature*; Eaton, *The Intellectual Journey of Thomas Berry*; Dalton and Simmons, *Ecotheology and the Practice of Hope*.

subtle form of *scientism*, even within these new creation myths.

From Ecological Equilibrium to Flux

In addition to what might be termed new ways of framing the debates, some scholars, including myself among them, want to be even more keenly aware of shifts in the basic philosophy of ecological science, as briefly noted in Box 2 above. Up until about ninety years ago, the science of ecology was thought of as a system in a stable equilibrium state, with humans understood as in some sense apart from that ecosystem ecology. But over time the idea of flux and unstable equilibrium became more prominent. Humans are now just as much a part of the ecosystem as any other creature, except with one vitally important difference. Humans are at least potentially more self-aware of their impacts on the planet than other creatures. An early focus on preserving biodiversity has given way to more complex forms of environmental studies that include human ecology as well.

From Creation Theology to a New Hermeneutics to New Constructive Theologies

A further trend that I would like to flag up in this overview chapter is the way the discussion has shifted from early concentration on developing new theologies of creation, to a much wider brief on interpreting scriptural material and systematic theology. The new theologies of creation of course had their place, and Jürgen Moltmann was perhaps one of the most successful and influential advocates of creation theologies that took into account ecological issues. This work is still ongoing, but in addition to that move, more radical ways of approaching biblical studies began to surface. Biblical scholarship was formulated in a way that deliberately sought to view biblical texts through a different lens, namely, the lens of ecological concern. Finding new approaches to *all aspects* of systematic theology is rather more recent.[23] Theologies of creation and to some extent Trinitarian thinking have always been prominent. But now rather greater attention is being paid to a fresh discussion of anthropology, the Spirit, pneumatology, Christology, the future of creation, eschatology, the problem of ecological and evolutionary suffering, theodicy, and so on.

This new systematic approach is "chastened," as it were, by a more deliberate attempt to engage the practical issues of concern in a way that, arguably, the earlier attempts at theology did not. A dilemma arises in such discussion. How far does reading God through an ecological lens really provide a practical basis for dealing with severe and complex ecological problems? Just because our understanding of God incorporates ecological dimensions, does this necessarily help us solve problems associated with environmental devastation? It might even increase human guilt, but not much else. On the other hand, if systematic theology routinely ignores ecological issues, it could reinforce the idea that ecology does not have much to do with theology and so can be dispensed with as a serious matter of concern for the church.

23. Conradie et al., *Christian Faith and the Earth.*

From Environmental Ethics to Creaturely Ethics

Pioneers of a secular approach to environmental ethics took a point of view on the natural world where the greatest concern went to protecting systems, especially ecosystems, from environmental harms, along with maintaining species diversity, as noted earlier in this chapter. Another movement focused on the specific needs of animals and their treatment, with an allied discussion of animal welfare, animal liberation, and animal rights. Environmental ethics assumed that moral priority had to be given to adequate functions of ecosystems. Animal rights, on the other hand, extended questions about justice for humans to other individual animals, mostly sentient animals. Tensions in these perspectives are obvious, but more recently scholars have pressed for a coalition so that human responsibilities toward creatures and their living environment are taken into account.[24] The latter brings with it an important allied literature of animal studies and the environmental humanities, both of which can serve to invigorate environmental ethics in new and interesting ways.

From Political Advocacy to Public Theology

This trend is worth naming as it represents the tendency, on the one hand, for social and political discussion about the environment to exclude religious beliefs, and on the other hand, for ecotheologians to ignore political and social concerns. Those ecotheologians who have faithfully represented a theological position in the midst of secular debates on environmental problems include Protestant

24. Deane-Drummond and Clough, *Creaturely Theology*; Deane-Drummond et al., eds., *Animals as Religious Subjects*.

authors such as Peter Scott and Michael Northcott.[25] But there are other trends afoot in this area as well from a Roman Catholic perspective, not least the broadening out of concerns for those who are poor. Liberation theologians are known for their concentrated critique on economic and structural issues. Leonardo Boff's incorporation of Gaia as a framework for discussion will only be attractive to those who are convinced by this approach.[26] Public theology, that is, theology that is deliberately intended to contribute to public debate and discussion, also has an important role to play in addressing environmental problems.[27]

REFLECTION

1. Do you think that agrarianism helps recover the sense of connection with the earth and ourselves as creatures? Why? What are the benefits and limitations of this approach?
2. How far might a rewriting of systematic theology along ecological lines convince you of the importance of ecological concerns for Christian theology? If the answer is yes, then which areas of systematic thought would you begin with and why? If the answer is no, what are the alternative strategies?

25. Northcott, *The Environment*; Scott, *A Political Theology*.

26. Boff, *Ecology and Liberation*.

27. Deane-Drummond and Bedford-Strohm, eds., *Religion and Ecology*.

2

ECOLOGICAL BIBLICAL HERMENEUTICS

This chapter gives insight into how to read the Bible ecologically from the perspective of the Earth, examines controversies over the legitimacy of such an approach, and provides a discussion of significant texts, including, for example, creation accounts from Genesis and the Wisdom literature.

The millennials, those born between 1981 and 2000, are, according the most recent pew research, more likely to trust institutions than their predecessors born after World War II, also called the "baby boomers." The millennials are important, since they are poised to make a substantial contribution to patterns of consumption and production in the next half century. The Bible has authority across all different Christian institutions, though the particular weight of that authority will be higher for Protestant denominations. The Catholic Church has encouraged lay access to the Bible ever since the Second Vatican Council permitted a much closer reading of Scriptures.[1] Hence, for all these reasons, any version of ecotheology needs to consider how far and to what extent the Bible might support, or even work against, a specific concern for the natural environment.

Part of the difficulty is that the Bible can be interpreted in different ways that are incompatible with each other. For example, when the Bible talks about the "world" does it mean the whole natural cosmos, or just human culture? Ecotheologians have set out to read Scripture either directly or indirectly, in terms of its relevance for current ecological problems, even though the specific problems faced in ancient times were different.[2] This parts company from *eisegesis*, a reading *into* the text something that is not there at all arising from one's own biases and presuppositions, but is rather a new kind of interpretation of the text or *exegesis* that appreciates the context of the writers in a new way. Each text also has its own *reception history* in given communities, that is, how it has been interpreted across different generations.

1. One of the hallmarks of the Second Vatican Ecumenical Council of the Roman Catholic Church is the attention it gave to the authority of scripture. *Dei Verbum* (*The Word of God*), published in 1965, reinforced the importance of Scripture and called for a biblical revival in the life of the Church. It was only really after the Council that Roman Catholic scholars were given official ecclesial blessing to use the tools of biblical scholarship such as historical and literary criticism.

2. For a scholarly overview see Horrell, *The Bible and Environment*.

BOX 1: AN AGRARIAN READING OF THE BIBLE

Ellen Davis, like many other North American scholars, is attracted to the idea of agrarianism. She claims "an agrarian reading of the Bible is not an exercise in nostalgia, although it is in significant part a work of memory, of imagination anchored (not mired) in the past." A good example of how that might play out in practice is Leviticus 19, where there are eighteen directives following the initial demand for holiness. So, "You shall act lovingly to your neighbor, as {to} yourself" (19:18) is included along with the agricultural "You shall not mate your animals in two kinds nor seed your field in two kinds" (19:19). Davis believes these two texts are deliberately put together, so the string of concrete suggestions shows us by analogy what holiness looks like "on the ground." The prohibition against mixing kinds may be related to the fear that mixtures trespassed on the holiness of mixtures; for example the cherubim was a hybrid who guarded the holy arc of Israel's sanctuary. She resists the idea that this is some sort of literal endorsement of monoculture, since the Leviticus writer deliberately wanted to provoke careful thought about how we trespass on the holy in the way we practice agriculture. For her, the equivalent scenario today would be industrialized agriculture, the use of genetically modified crops, or those that undermine local farming practices.

Davis, *Scripture, Culture and Agriculture*, 4, 85.

An agrarian reading of Scripture, in its focus on the agricultural and cultural background of the texts, helps to bring those Scriptures alive as we imagine the natural and living environment in which these ancient peoples lived and worked.

But it also points to other aspects of ecotheology that are important in more general terms. These are:

1. A much greater attention to biblical stories about the relationships between God, humanity, and creation, rather than just focusing on human history.

2. A consideration of creatures other than humans as an integral part of that creation.
3. Special attention on the wisdom literature, such as the Proverbs, the Psalms, the book of Wisdom, or even the book of Job, which develops its own unique theology of creation.
4. A rereading of the ministry of Jesus that pays attention to the way he relates to the natural world around him.
5. A recovery of the true meaning of apocalyptic literature, which has sometimes been misinterpreted to imply a rejection of the natural world.

BOX 2: NATURE OR CREATION?

Why do biblical scholars more often than not use the term *creation* rather than *nature*? There are a number of reasons for such usage, most important perhaps is that for Christians who believe that God is a divine Creator—drawing particularly on the first book of the Bible, Genesis—speech about creation and the idea that the earth is a gift of God makes a great deal of sense. Nature, on the other hand, has multiple meanings, but is most commonly thought of as that part of the world that is not human and can be understood through knowledge arising from the natural sciences. Those outside the Christian community would not readily appreciate a presupposed understanding of the world as that created by God. Of course, many authors will deliberately bridge both discussions. One way to understand this is that there are two sides to the same coin, a reading of the natural world as nature, and a reading as divine gift or creation. This makes a scientific and theological understanding of the natural world around us fully compatible and therefore undercuts the supposition that creation is opposed to scientific ways of knowing the world.

DOES THE BIBLE PROMOTE CARE FOR CREATION?[3]

The theme that the Bible provides impetus for human care of the earth is one of the most rehearsed in the literature. At least one reason for this profusion stems from reactions to American historian Lynn White, who in 1967, early on in the environmental movement, claimed that the Judeo/Christian tradition was responsible for causing the environmental crisis.[4] While rhetoric of "an ecologic crisis" has generally now faded, more common is the claim that dangerous and irreversible climate change is taking place, and is perhaps even more threatening. White believed that the Bible, specifically Genesis 1:28, endorsed domination of the earth. More bluntly, he pointed to the way Christian natural theology endorsed the rise of science and technology that then, he suggests, subsequently became out of control. It thus led to Christianity's "huge burden of guilt."[5] He is critical, in particular, of the development of attitudes that separated the divine and human from the natural world through its emphasis on human and divine power and transcendence, and the natural world's desacralization and subsequent exploitation. He has influenced a generation of both secular and Christian scholars, with many of the former dismissing Christianity as a valid resource for responsible ecological action. From a historical perspective, blaming Christianity alone for the rise in science and technology and linking this with ecological stress is far too simplistic, but White's rhetoric struck a chord with many who saw in Christianity,

3. Some aspects of this chapter are drawn from my earlier book, Deane-Drummond, *Ecotheology*, 81–98.

4. White, "The Historical Roots."

5. Ibid., 1206.

compared with religious traditions such as Buddhism for example, an over emphasis on human supremacy.

Christian and Jewish scholars have, accordingly, leapt to the defense of the Biblical story in most cases, arguing that dominion does not mean domination; rather it means humans taking responsibility for the earth as stewards and viceroys of that creation. The Bible speaks of human distinctiveness from other creatures; for only humans are made in the image of God (Gen 1:26) and only humans are called to name the animals (Gen 2:19–20). But there are also considerable lines of continuity among all creatures expressed in the biblical account as well. For example, God created humans on the same day as the other animals, and Adam means quite literally "of the earth," while Eve means "daughter of life." The particular vocation of humans is to "subdue" and "have dominion over" the earth, but if this is seen in the light of humanity's role as divine image bearing, then such terms could not mean "exploitation," but rather, careful service for the earth. The biblical account states that humanity became a living being through the breath of God (Gen 2:7), hence, ensoulment emerged *from* earth material, rather than being *added to* that material in a dualistic way. Other scholars have found in the Bible a source of positive directives for care for the earth.

BOX 3: SEVEN BIBLICAL PRINCIPLES OF STEWARDSHIP

Environmentalist and Christian evangelical writer Calvin B. DeWitt identifies the following seven principles as emerging from this general theme:

1. *We must keep the creation as God keeps us.* Human *earth-keeping* (Gen 2:15) mirrors the providence of God in keeping human beings (Num 6:24–6). Dominion

is exercised after the pattern of Christ, so that humanity joins with the Creator in caring for the land (Deut 11:11–12 and Deut 17:18–20).

2. *We must be disciples of the Last Adam, not the first Adam.* Just as in Christ all things are reconciled (Col 1:19–20), so the human vocation is to participate in the restoration and reconciliation of all things.
3. *We must not press creation relentlessly, but provide for its Sabbath.* Exodus 20:8–11; 23:10–12 shows Sabbath rest applies to the land as well as animals and human beings.
4. *We may enjoy, but not destroy, the grace of God's good creation.* The tendency for human greed to destroy the *fruitfulness* of the earth is documented in the biblical accounts of human behavior, as in Ezek 34:18 and Deut 20:19; 22:6.
5. *We must seek first the kingdom, not self-interest.* The mandate for this comes from the Gospels, as in Matt 6:33.
6. *We must seek contentment as our great gain.* This means being content with the gifts that creation brings, rather than always grasping after more. There are therefore *limits* placed on humanity's role within creation. Paul's letters here give some encouragement, as in Heb 13:5 and Tim 6:6–21.
7. *We must not fail to act on what we know is right.* The marriage between belief and action needs to be fulfilled in stewardship practices. The need for a link between belief and action is a strong biblical theme, as in Ezek 33:30–32.

DeWitt, "Creation's Environmental Challenge," 65–67.

The notion that humanity might be capable of *improving* nature is also implicit in Calvin DeWitt's principle of restoration and reconciliation.

Environmental philosopher James Nash questioned *any* use of the Bible to support ecological practices such as defense of biodiversity, arguing that the Bible did not have such concerns, and the ethical imperatives in the Bible are ambiguous.[6] Most scholars, none the less, while taking note of Nash's concern for improper use of biblical texts, do believe that the Bible has the ability to make perennial ethical demands that are ecologically relevant.

THE BIBLE AND THE STATUS OF CREATION

Many authors have become unsatisfied with the biblically based theme of human stewardship. It leaves untouched the question of *why* we should care for creation, other than that humanity is commanded to do so by God. In other words, while creation care may be a consequence of a deeper understanding of the significance of creation, we need to learn to appreciate that significance first. Genesis speaks of all creation as good, as all creation is loved by God and gains its worth through this love. One way to avoid problems associated with the fact that at least some creatures seem to our own moral sensibilities to be far from good, is to interpret texts that refer to goodness to mean proper *functioning*, rather than making any explicit moral claim. Yet if we turn to one of the books in the Wisdom literature, Job (38–41), we find another picture of the relationship between humanity and the earth. Here the natural world in all its wildness is presented alongside humanity, who remains humbled before its savagery and ambiguity. Here God's knowledge outstrips human ingenuity and domestication of animals. Here God's authority reigns to support the needs of wildlife, rather than just that of human beings. Job makes more explicit the theme that God blesses all of creation even after humans have appeared. It also highlights an important strand that runs

6. Nash, "The Bible vs. Biodiversity."

through the literature of the Hebrew Bible, namely that of *blessing*, one that stresses the fecundity of God's creative activity, with an emphasis on celebration/joy in creation.

BOX 4: CREATION AND THE BOOK OF JOB

In the context of his extreme suffering, Job cries to God for explanation, and turns to listen to the suppressed voice of the earth (Job 12:7–9). Yet God's reply is one that expresses care for all of creation; wisdom exists in all wild creatures in a way that is quite independent of human society and its needs. Furthermore, while humans and earth creatures search for wisdom, God seemingly finds such wisdom while searching the depths of the earth—measuring the waters, weighing the wind, and ordering the storm (Job 28:23–27). It is not completely clear if God finds wisdom in the earth by accident or through a deliberate search, though God *acquires* such wisdom in the process of creating. God is now the *discoverer* of wisdom and looks to earth for this wisdom. Everything seems to have its *derek* or way, including light (Job 38:19), lightning (Job 38:24), and thunderstorms (Job 38:25). This seems to point to an ancient ecology that speaks of a belief that the earth is governed by regulated systems, whose principles follow natural wisdom, rather than through divine intervention as such.

Habel, "Where Can Wisdom Be Found?"

Wonder at the marvels of creation is also an attitude encouraged in the Biblical text. The transforming power of wonder is expressed beautifully in William Brown's recounting of what Job might have felt like after his experience of God in the whirlwind: "I imagine him laughing, like the ostrich, as he boldly reinvested himself in raising a family. I doubt Job ever roused himself early in the morning to offer sacrifices for fear his children had sinned. Gratuitous delight, rather than honor and fear, motivated Job's care for his children."[7]

The earliest wisdom literature seemed to be more concerned with justice in the human community, rather than centered on creation as such.[8] Yet, observation of the natural world is included as a basis for reflection in some proverbs, such as Prov 6:6–9. Here humans are exhorted to notice the "way of the ant". This could imply a way of finding truths in the natural order, a kind of implicit natural science, but it could also be an exemplar of a different kind of non-hierarchical social order that seems to be subversive to much of the wisdom literature that attributes wisdom to elite, royal, functions. In addition, in Proverbs 30 we find Agur comparing the world of nature and that of humanity—the way (*derek)* of an eagle in the sky, a snake on a rock, a ship on the high seas, and a man with a woman. Wisdom is also in the smallest and most insignificant of creatures—the ants, badgers, locusts, and lizards.

Alongside this affirmation of creation's worth, there is a responsive theme to that blessing, namely creation's praise of God. The theme finds its way into Biblical writings in both the Hebrew Bible and the New Testament, for example, Isa 42:10; Pss 19:1–4; 69:34; 96:11–12; 98:7–8; 103:22; 150:6; Phil 2:10; Rev 5:13. While these passages are metaphorical in that they claim that non-human creatures praise God in human language, they are certainly not expressions of "animism" or some kind of panpsychism that accords rationality to all living things, or even just poetic decoration.[9] The most likely explanation is that

7. Brown, *Wisdom's Wonder*, 135.

8. Habel, "Where Is the Voice of Earth."

9. Bauckham, *Bible and Ecology*.

in most cases the praise stems from creation's acknowledgement of it being created the way it is.

THE FUTURE OF CREATION

The possibility of deliverance or redemption of both people and land emerges from a theology of blessing, even though both are in some sense distinct from each other. The book of Exodus recounts a deliverance from a land of oppression to one that is "flowing with milk and honey," a figurative portrait of fecundity, but also where justice reigns on earth. Land is, moreover, not so much owned as given by divine gift that itself impacts particular practices.[10] The link between justice in the land and human community is an important facet of the prophetic tradition. The prophets after the exile declared that the restoration of the land required a return to integrity and justice. Isaiah's vision of *Shalom* was one replete with accounts of ecological harmony that is more obviously metaphorical.

In the New Testament, Paul's letter to the Colossians speaks of Christ as having redeemed all things; while this refers to *sin* in the case of humans, for creation at large, including humans, it means a future redemption from suffering, so it is through Christ that the future can be secured. Romans 8:19–23 also speaks of all of creation waiting its time of deliverance, along with the children of Israel. Creation is groaning now, but this groaning will not last indefinitely, for there will come a time when this groaning will be heard. This passage also links the suffering of creation with humanity's failure to act, so that the corollary is that appropriate human behavior is required

10. A point that Michael Northcott makes forcefully in *Place, Ecology and the Sacred*, 111–32.

for creation's flourishing. Where humanity has failed to act appropriately, the future of the whole universe is at stake. Yet right human action is that empowered by God's grace inspired by the Holy Spirit in humble service to that creation, rather than through an inordinate exercise of power over creation. The royal figure of Christ is paradoxically that which expresses the humility of the servant king.

The Sabbath rest on the seventh day is the ultimate goal of the creative process in Genesis, so it is the Sabbath rather than humanity that is the "crown of creation." The rest of God in creation stands in stark contrast to the empty waste and brooding uncertainty that was there before the first creative acts. The state of the earth before the creative acts of God described in Genesis 1:2 is bare and uninhabited—and it seems that there is no need to assume some kind of "chaos" reigned prior to these acts. The rest enjoyed by God following active creation is not so much passive rest as *active appreciation*, and it puts the creative work of the previous six days in perspective. Now the holy is attributed to time, rather than space.

Norman Wirzba has suggested that living the Sabbath applies across a broad range of activities, including taking responsibility for the natural environment.[11] He believes that our cultural detachment from our creaturely context means that we are no longer able to respond to the gift of God in creation. He also draws on Psalm 65 as a way of illustrating the intimate involvement of God in the fecundity of creation. The Sabbath encourages a different way of being, so that "the Sabbath will not allow us to separate ourselves from the rest of creation and the creation from God. For as soon as we make our separation, we condemn ourselves to loneliness and creation to

11. Wirzba, *Living the Sabbath*, 142–53.

violence."[12] Exodus 23:10–11 and Lev 25:3–7 both speak of the importance of the sabbatical year of rest for people, their animals, and the land.

The Sabbath points to the ultimate purpose that God intends for all creation, namely to give glory to God by being what God has intended it to be through God's wisdom. Perhaps the greatest challenge in the development of eco-practice is setting aside a particular time for just being with the natural world and with God, in active rest understood as appreciation. Creation is not just "good," it is also declared holy. It is in such contexts that we can, perhaps, begin to hear the silent "voice" of creation and its demands.

THE BIBLE FROM AN EARTH PERSPECTIVE

While Biblical hermeneutics at one time attempted to distance itself from the assumptions of the interpreter, contemporary biblical scholars are more likely to admit that such a stance is virtually impossible, even if unacknowledged or unconscious. Indeed, rather than making any claim for neutrality, we need to read the Bible through particular lenses in order to highlight different facets of its meaning. The Earth Bible project[13] invites a particular reading of the Bible in the light of particular principles; it claims that its readings open up horizons that have been ignored in the past. This is not the same as reading particular meanings into the text, for it allows for literary and cultural analysis as well. Some have argued that the Earth Bible principles used

12. Ibid., 145.

13 The Earth Bible is an international research project that focuses readings of major sections of the Bible according to ecojustice principles and was launched by Sheffield Academic Press.

are not open enough to allow alternative interpretations, but it is also apparent that this strategy is a corrective one. The principles themselves were arrived at in consultation with ecologists as well as biblical scholars; it therefore represents a listening to the natural world that is in tune with the biblical wisdom tradition.

BOX 5: ECO-JUSTICE PRINCIPLES OF THE EARTH BIBLE

Norman Habel has proposed six eco-justice principles, all of which have significance in relation to interpretation of the Bible:

1. The first principle of *intrinsic worth* relates to the worth of creatures.
2. The second principle of *interconnectedness* is one that is universally familiar to ecologists and environmentalists.
3. The third principle of *voice* claims that the earth is capable of raising its voice in celebration and against injustice, so viewing the Earth in *kinship* with rather than in alienation from humanity.
4. The fourth principle of *purpose* claims that the universe, the earth and all its components are part of a dynamic cosmic design, where each contributes to that God-given purpose.
5. The fifth principle of mutual *custodianship* reflects on the role of humans in relation to the earth. Humans should think of themselves as guests on planet Earth, custodians of their host Earth.
6. The sixth principle of *resistance* claims that the earth and its components actively resist those injustices imposed by humans.

The Earth Bible Team, "Guiding Ecojustice Principles."

Such principles become easier to understand with some specific illustrations from the biblical text. The principle of *intrinsic* worth is relevant to an interpretation of 1 Tim 4:1–5 that asserts the goodness of the earth over against others who might claim otherwise. This pastoral letter was challenging the claims of those who advocated ascetic practices such as forbidding marriage and abstinence from food. For Timothy food is to be received with thanksgiving, and everything that God has created is good (1 Tim 4:3–4). First Timothy 4:5 states that *all of creation* is sanctified or consecrated, that is, made holy. This contrasts with more traditional uses of sanctification as applicable just to the human community.

The principle of *interconnectedness* becomes relevant for the interpretation of the narrative in Genesis 4. Arguably the third player in the well-known narrative of the story of Cain and Yahweh is actually the earth itself. Cain, who worked on the land, wanted a blessing from Yahweh for his labor, but he failed to secure this blessing. His response to his brother Abel, who kept sheep, was to murder him. Yahweh then claims that the voice of Abel's blood *cries out from the ground* (Gen 4:10). Cain can no longer live off the soil, he becomes totally alienated from the cursed earth, and is forced to be a wanderer. Vengeance comes through the flood, with Noah, named as a man of the soil, one who had a right relationship with the earth, as the one singled out for deliverance. Noah is the first to plant a vineyard after the flood, again affirming the crucial importance of right relationship with the land.

The principle of *voice* is also illustrated in the above account, where the earth is spoken of as metaphorically "crying out" to Yahweh for the blood of Abel. There are other passages that more explicitly speak of the lament of the land, such as Jeremiah 12. The lament reaches its climax in 12:4, an appeal on behalf of all creatures. The desolate land mourns to God (12:11) and God responds (16:19), which indicates a direct relationship with the land. Jeremiah claims that moral order affects the order in creation. Divine anger is partly responsible for environmental degradation in Jer 4:23–26. Such anger is provoked by Israel's infidelity, but alongside this anger is grief, touching God's heart (Jer 12:7, 11). God's own possession or "heritage," which links both people and land, is forsaken and therefore no longer the subject of God's blessing. Although the land belongs to God, God gives responsibility for it to others, revealing a divine vulnerability. God now calls upon wild animals to be instruments of judgment, hinting at the eco-justice principle of *resistance* (12:9). The final vision that Jeremiah presents is one of hope of salvation that is universally inclusive of all peoples and land.

The principle of *purpose* is illustrated in Rom 8:18–22. Creation is subject to futility against its will, so a hope in reversal is retained where human beings would once again find favor with God. This passage assumes Paul believed in a traditional interpretation of Genesis 1 where humanity is set over against the rest of the created order. The subjugation of the earth spoken of in Romans 8 could be viewed as connected with the sin of Adam discussed in Romans 5. When viewed in the light of Christ, human beings respond to God's grace and act accordingly. The depiction of creation suffering is that of birth pangs, promising a positive future for the earth.

The fifth principle of *custodianship* is rather more difficult to find. Psalm 8 could be read as the opposite, an apology for human (particularly male) domination of the earth. Norman Habel finds in Genesis 1 a story that begins with the earth and its flourishing, but then shifts to a story where the earth takes a secondary role under *adam*, who has power over all life. Humans are expected to subdue (*kabash*) the earth, used in other contexts to express forceful subjugation. Habel believes that we cannot escape this negation. Rather than try and rehabilitate difficult texts that speak of human domination, Habel suggests something far more radical, namely giving the first half of the story about the earth far greater prominence. The consequence of such a reading would be a deeper sense of custodianship and kinship with the earth, even if the principle of custodianship rarely finds its expression in the biblical text as such.

These newer readings of the Bible are controversial inasmuch as they depend on looking at the text with a specific question in mind. Critics believe that finding evidence for specific ecological threads in Scripture breaks with the long history of interpretation. An alternative is to argue that the biblical stories support a world view or ethos that is relevant for environmental action today,[14] rather than claiming direct relevance. Others, myself included, have wanted to include ecologically relevant virtues such as faith, hope, and love, along with justice, wisdom, temperance, and humility as an aspect of what biblical traditions can affirm.[15] Supporters of explicit ecological readings argue that naming those lenses is no different from traditional biblical scholarship that is always grounded in a particular context and place.

14. Barker, *Creation*, 1–33.

15. Deane-Drummond, "The Bible and Environmental Ethics"; see chapter 7 for further discussion on virtues.

REFLECTION

1. How would you defend Christianity against the charge that its human centered (anthropocentric) ethos has contributed to environmental harms?
2. Compare the seven principles of DeWitt with the Earth Bible's eco-justice principles. Which do you find most convincing and why?
3. Do you think that the Bible can still make moral demands today? Why?
4. What biblical themes do you think are most relevant for encouraging ecological activism? Why?

3

NEW ECOTHEOLOGIES OF LIBERATION

This chapter deals with the context of liberation theology and its engagement with global issues of poverty and how this impinges on environmental concern. The work of Leonardo Boff will be discussed, and a comparison made with ecofeminist approaches from a range of perspectives.

Have you ever walked through an underpass and seen the figure of a homeless man or woman, crumpled against a wall with a dirty cardboard poster begging for money? Since I arrived in Indiana in 2011, I've seen anguished stories written out in semi-literate fashion in jagged print, telling of their babies with no food, sickness of all kinds, war veterans who felt they were left to rot, and so on. Experiences like this are particularly common in the United States, which boasts of being one of the richest nations of the world, but also has a poor record in welfare provided by the state. Liberation theologians are those who want to side with the marginalized in society, either within rich nations, or in the poorest regions of the world. Most importantly, for these theologians, dire poverty is not just a situation that happens to some people because they are out of luck, have developed bad addictive habits, or are mentally sick, but it is related to the social and political climate of that community. Liberation theology was birthed in the experience of those who witnessed the suffering of those oppressed by various totalitarian regimes and other unjust political structures in the poorest nations of the world, particularly Latin America. The question then becomes: how can that poverty be addressed? A strong criticism of the "development" rhetoric of the Western world is that it amounts to importing particular cultural values. Initially liberation theology focused more specifically on human survival and flourishing in the particular contexts of oppressive regimes, but the association of the devastation of land with poverty meant that the cry of the poor also joined with a cry of the earth.

POVERTY, "DEVELOPMENT," AND ECOLOGY

In the 1940s the most powerful voices were in favor of a particular model of development known as "modernization." So-called "underdeveloped" nations were required to find ways of becoming productive in economic terms, boosted by input from so called "developed" richer nations. The assumption was that such a development strategy would pave the way for stable democratic societies to emerge. André Frank challenged the modernization thesis through an alternative known as "dependency theory." He attacked capitalism and trickledown theory as a means for development in Latin America, arguing instead that foreign monopolies or domestic elites siphoned off profits, leading to overall stagnation. Gustavo Gutiérrez argued that the term "liberation" should be used, rather than "development," for it avoided the pejorative connotations associated with the latter, and also had the advantage of linking up with more biblical themes.[1] For him, revolutionary forms of socialism are the only way to break the bonds of an unjust society, epitomized in forms of class struggle at a local, national, and international level. Even though dependency theory attacked modernization in principle, how was this realized in concrete solutions to political and social dilemmas?

Historically the divide between development and environmental concern was not simply about choosing between people or pandas, but a radical difference in approach to social issues. While those on the "red" end of the spectrum tended to view capitalism as the primary source of problems, those on the "green" end generally viewed technological culture as something to be vigorously opposed. Some of those concerned with environmental issues, such as The Environmental Fund, a group founded in the 1970s, even advocated a form of ecofascism. Pessimistic about the prospect of improvement for peoples in the poorer nations of the world, they were prepared to advocate more extreme curtailment of population growth. From about the 1980s onwards there was a wider recognition that economic growth had to go hand in hand with ecological sensitivity in order to preserve the long-term future of peoples. Alternatives included the idea of "endogenous" development, which is an awareness

1. Gutiérrez, *A Theology of Liberation*.

of local and global limits, or in Catholic social thought, integral development.

Gradually the rhetoric of sustainability and sustainable development emerged. Unfortunately this does not necessarily refer to encouraging styles of living that are commensurate with the earth's carrying capacity. Sometimes it is the sustainability of human societies that is being referred to in terms of population, consumption, resource use, and pollution, including the obligation to future human generations. More insidious, perhaps, is the imposition of highly industrialist farming methods on fragile communities, which are then vulnerable to erosion and loss of livelihoods.

BOX 1: CASE STUDY: KINAPAT FARMING PRACTICES

When loggers first cleared the forests, early settlers moved in and formed a small hillside village of Kinapat, the Philippines. Initially, one hectare yielded about sixty to eighty sacks of maize. By the 1980s, some twenty years later, the yield had dropped by 83%, and more than half of the working population had moved on to alternative, more fertile lands, only to repeat the cycle. Official agricultural officers believed that yields were low because of outdated farming methods. They recommended high yielding varieties alongside chemical fertilizers and pesticides. The best land had always been given to large agricultural business plantations, reliant on high yielding varieties and input of chemical fertilizers. This left the poor the only option of farming on marginal hillside farms following clearance of the rainforest. Farming is even more precarious in these conditions, with rainfall drastically reduced and crops vulnerable to pests, drought, or flash floods. Once the soil is washed away it is carried down to the sea, silting up coral reefs, and destroying habitat for fish stocks. The recommendation to use high yield varieties and other intensive farming methods was impossible for economic reasons. Local farmers in Kinapat believed that yields were not low because they refused intensive farming methods; rather the crucial change was the gradual loss of fertility of the soil. Three farmers decided to undertake a participatory research survey into the reasons for low yields. They documented evidence for much higher productivity when they first moved to the site. If the soil was originally productive, then their traditional farming methods had been adequate in the past, hence it was the loss in soil productivity that accounted for the low yields. They believed that a simple alternative known as sloping agricultural land technology (SALT) would prevent soil erosion and help rehabilitate viable farms. Some farmers worried that they might be driven off the land for other reasons, as they did not hold official titles to the land. However, in spite of these reservations the farmers banded together to form the Kinapat SALT Farmer's Association, believing that simply moving on to other sites would be disruptive to their communities. They also held out the hope that they could apply and obtain titles for their land, which materialized in due course.

White and Tiongco, *Doing Theology and Development*, 154–58.

ECOLOGY AND LIBERATION

Leonardo Boff is, arguably, one of the foremost liberation theologians who has actively sought to integrate a theology of liberation with environmental concerns.[2] He writes from his experience of the threatened fragile area of rainforest in the Amazon region of his native Brazil. He begins his book entitled *Ecology and Liberation* with a rehearsal of the need for greater awareness of global ecological issues. He is particularly critical of forms of eco-politics that focus on organic, conservation, or additive free products while failing to take into account basic human needs and global environmental concerns. The task ahead is to

2. Boff, *Ecology and Liberation.*

heal the "broken alliance" between humanity and nature, individuals and community. His agenda for a new global political economy lays out the following elements:

a. A minimum of humanization that honors freedom and respects individual human identity. Basic human needs should be met, including, for example, requirements for basic food, shelter, and health care.
b. Citizenship that is inclusive of all people in their diversity.
c. Equity that is a greater realization of political equality.
d. Human and ecological welfare. The goal should be an enhanced quality of life, not simply an improvement of goods and services.
e. Respect for cultural differences.
f. Cultural reciprocity and complementarity.

He believes that modern civilization has as its organizing axis not life, or the wonder of life, but its own power and means for more power and domination. The common ground between liberation theology and radical ecology is that they both:

> [S]tart from two bleeding wounds. The wound of poverty breaks the social fabric of millions and millions of poor people around the world. The other wound, systematic assault of the earth, breaks down the balance of the planet, which is under threat from the plundering of development as practiced by contemporary global societies.[3]

3. Boff, *Cry of the Earth*, 104.

He believes that not only are both accounts starting in awareness of suffering, but also both accounts of suffering are rooted in a similar cause, namely a social system that encourages accumulation and exploitation of both people and the plundering of natural world. He wavers between a radical liberation theology that focuses on the liberation of human beings and a radical ecology that gives the highest priority to protection of the earth, given that without the earth there could be no human life anyway.

Gradually liberation theology has also moderated under the influence of indigenous theology through its attention to the importance of base communities. Indigenous cultures seek to stress primarily *identification with the land*, rather than radical economic critique of capitalism through socialist ideology, though that has started to shift with politically active protest groups forming, such as that over the Dakota pipeline. Indigenous traditions are also important inasmuch as they put due emphasis on a theology of place.

Simon Schama in *Landscape and Memory* explores the extent to which historically those in Europe and North America have been sensitive to the indissoluble connections between people and place.[4] He emphasizes that we cannot separate the natural world from human culture; the landscape is an expression of such interaction. Such a view challenges the perception that it is possible for human existence simply to tune into what is inherent in the natural world without alteration, a view that seems implicit in indigenous traditions. This perspective is also rather different from the romantic American tradition of wilderness represented in nineteenth century American environmentalists such as Henry David Thoreau

4. Schama, *Landscape and Memory*.

(1817–1862), John Muir (1838–1918), Walt Whitman (1819–1892), and Susan Fenimore Cooper (1813–1894) that are still very influential among American ecotheologians. It is also possible to learn from this historical work that there are resources buried in the Western tradition that are more than the caricature of exploitation, capitalism, and aggression implied by liberation theologies.

BOX 2: INTEGRATING AFRICAN TRADITIONS

The alliance of African Independent Churches (AIC) have sought to weave together insights from indigenous African cultures and Christian faith. These efforts function as a case study where indigenous religious views have been integrated into a liberation approach from grass roots practice. One hundred and fifty of these churches, inspired by the work of the Zimbabwean Institute of Religious Research and Ecological Conservation (ZIRRCON) have formed an Association of African Earthkeeping Churches (AAEC) and also collaborated with groups from a traditional African religious perspective, namely the Association of Zimbabwean Traditionalist Ecologists (AZTREC), comprising chiefs, headmen, and spiritual mediums. Together they have succeeded in planting between three and four million trees in Zimbabwe, incorporating tree planting into religious rituals. Religious leaders in the AAEC viewed such tree planting by poor and relatively under-privileged members of society as a sacred mission, fulfilling a vocation to heal the earth. This task is woven into the more traditional evangelical task of conversion to Christian faith. The portrayal of the church as guardians of creation links with the traditional Shona cosmology, where founder ancestors of tribes and clans are referred to as guardians of the land, protectors of the holy groves and sanctuaries. In association with Christian faith, Christ becomes the universal guardian of the land, prefigured in that of ancestors, yet his Spirit is understood to hold sway over Shona guardians. Those who willfully destroy the earth are labelled as "wizards," in parallel with those opposed to the liberation struggle prior to the independence of Zimbabwe. The first president of the AAEC,

Bishop Machokoto, named the connection between the two in unambiguous terms, suggesting that "there is absolutely no doubt about the connection between our war of the trees and the former liberation struggle, *chimurenga*" (545).

Daneel, "Earthkeeping Churches," 531–52.

ECOFEMINIST THEOLOGY

Feminists have taken a leading role in active practical concern for environmental questions and issues. Not only are women historically associated with nature, but also cultural oppressions have served to inhibit women through dualisms inherent in patriarchy. Environmental injustice globally falls disproportionally on women.[5] Some feminists have described ecofeminism as the "third wave" of feminism. In a manner similar to liberation theology, feminist thinking puts most emphasis on *praxis*, which is the active concern for the way more theoretical concepts feed into practices and *vice versa*. Political positions within ecofeminism vary considerably, from more conservative through to more radical revisioning of politics. Heather Eaton compares ecofeminism with an intersection of different roadways, a meeting place of activists, environmentalists, and feminists, along with local and national groups.[6]

WOMEN AND NATURE

Many ecofeminists seek to find an alternative way of expressing spirituality that, for traditional thinkers at least,

5. For further discussion of gender issues see Cahill, "The Environment, the Common Good and Women's Participation."

6. Eaton, *Introducing Ecofeminist Theologies*, 3.

is not necessarily acceptable from a Christian theological perspective. Given the number of examples of the way women are portrayed negatively in the Biblical record, some more radical ecofeminists have given up finding Scripture as inspirational. Susan Griffin, for example, in her book *Women and Nature: The Roaring Inside Her* traces the dualisms between soul-flesh, mind-feelings, and culture-nature as the outcome of men confronted with the terror of mortality.[7] Instead of facing such mortality, she suggests that men oppress women and nature. Hence for her the solution is to identify with the earth in its mortality, the voice of nature that is joined with those of women becomes embodied and impassioned.

Rosemary Radford Ruether, more conscious of her Roman Catholic roots, rejects those aspects of a goddess "thealogy" that promotes religious practices without taking into account economic and social structures that have led to particular patterns of oppression. She is drawn to the new creation story of Thomas Berry, but weaves it into her version of a way of thinking about the earth as an interacting organism.[8] For her, Gaia represents the sacramental tradition, particularly, the cosmological presence of the divine in the natural world. Resurrection is interpreted in terms of the continuation of our bodily matter in future life forms on earth, leading to what she terms a "spirituality of recycling" that is only possible once humanity has experienced a "deep conversion of consciousness."[9] Heather Eaton more explicitly rejects any Christian salvific accounts as for her

7. Griffin, *Women and Nature*.

8. Ruether, *Gaia and God*.

9. Ruether, *Introducing Redemption*, 119.

such accounts are reinforcing a rejection of this world, rather than its affirmation.[10]

Mary Grey turns less to the covenantal and sacramental spirituality that is at the background of Ruether's analysis, and more to biblical prophetic themes as sources of inspiration in what she calls an "outrageous pursuit of hope."[11] Like Ruether she situates her discussion in cultural analysis, highlighting the culture of consumerism as of critical importance in fostering a culture where wants become needs, indirectly leading to exploitative attitudes toward the environment. Instead, she argues for a prophetic vision of flourishing drawn from Isaiah, one that is inclusive of both people and planet, and one that does not split ecology from social justice.

For Grey ecofeminist spirituality, like liberation theology more generally, arises from the margins and out of the concrete concerns linking the devastation of the earth and the suffering of vulnerable people.[12] She also argues not just for an alternative spirituality that is sensitive to the needs of the earth, but that also takes particular cognizance of the threats posed to human societies through globalization.[13] In particular, she believes that globalization poses as an effective spirituality, a misplaced desire of the heart; this needs to be tackled by providing an alternative, one that re-educates desire. She uses diverse ways of communicating that appeal to the person as a whole, such as reasoning combined with storytelling, along with

10. Eaton, "Epilogue."

11. Grey, *The Outrageous Pursuit of Hope.*

12. Grey, *Sacred Longings.*

13. Ideas related to the critique of consumerism and globalization are also argued for by Pope Francis in *Laudato Si*, though they are shorn of any explicit ecofeminist association.

parables and myths that are deliberately interlaced so as to appeal to the imagination and emotions.

BOX 3: COMMITMENT TO PRAXIS

Ecofeminist theologians, like liberation theologians, are committed to working on the interrelationships between theory and practice, or praxis. Mary Grey, for example, is the founding member of a non-government organization (NGO) that builds wells in India, and this practical experience filters into the way she writes with both passion and conviction about the experience of Indian women in the lowest castes and their struggle for human dignity. In particular, she challenges her readers who are likely to be mostly Western, to embrace a way of renunciation, simplicity, and sacrifice. In offering an alternative spirituality she succeeds in appealing to those from a variety of religious perspectives. Like many other ecofeminist writings she is more concerned with engaging practical contexts and the challenges these pose, rather than more specific theological analyses that tend to be more theoretical.

Wells for India, https://www.wellsforindia.org.

Should women be identified with the natural world or not? If they are, this seems to provide an excuse for oppression; if they are not, it seems to buy into the philosophical dualism between people and the natural world that they reject. Stacy Alaimo expresses this poignantly: "Speaking for nature can be yet another form of silencing, as nature is blanketed in the human voice. Even a feminist voice is nonetheless human: representing cows as ruminating over the beauty of the mother-child bond no doubt says more about cultural feminism than it does about cows."[14]

14. Alaimo, *Undomesticated Ground*, 182.

ECOFEMINIST LIBERATION THEOLOGY

Grace Jantzen's *God's World, God's Body* and Sallie McFague's *The Body of God* have been influential not just as ecofeminist theologies, but also within debates in systematic theology as well.[15] Jantzen suggests that just as humans are embodied, rather than existing as detached souls and bodies, so God too is embodied in the world, and God's transcendence is analogous to that of human beings, hence parting company from the traditional separation between God and the world. Jantzen believes that it is the universe as such that is expressive of the intentions and will of God. God as embodiment is costly to God in that God's power is self-limited by the desire to love. While notions such as the self-limitation of God's power are not unique to ecofeminism, Jantzen takes this up in a new way by incorporating the idea of God as feminine divine.[16] Like many other feminist writers, she is strongly critical of dualistic tendencies that she finds in Western culture, believing that under such dualism is a desire to control, leading to controlling attitudes not just toward women and nature, but also toward sexuality, feelings, and other races. She believes that the fear that underlies this dualism is fear of the body. She is explicit in her celebration of pantheism, the identification of God with the world. So that "instead of mastery over the earth which is rapidly bringing about its destruction, there would be reverence and sensitivity; instead of seeing domination as godlike we would recognize it as utterly contradictory to divinity."[17] However, would such a change in attitude

15. Jantzen, *God's World, God's Body*; McFague, *The Body of God*.

16. Jantzen, *Becoming Divine*.

17. Ibid., 269.

necessarily follow from changes in beliefs about God in the way she suggests? For her transcendence is reinterpreted through the idea of becoming, expressed in the feminine divine. She believes that those who reject such a view express a fear of being swallowed up in the maternal womb, associated with a loss of boundaries. But are not some boundaries helpful in maintaining distinctions?

BOX 4: THE EARTH AS GOD'S BODY

Sallie McFague, like Jantzen, also argues for an embodied model of God that has implications for the way we think about the earth. Instead of viewing the earth through an "arrogant eye" as like a machine that we then seek to control she suggests that we need to pay attention to the earth, come in tune with that Earth and become conscious of its vibrant subjectivity. Although she is conscious of the earth sciences and evolutionary theory, she undercuts their credibility by giving the earth as such subjectivity. For her sin becomes a refusal to accept our place on the earth. The planet is a reflection of "God's back" and the idea of the earth as God's body is deliberately metaphorical, so that we are "invited to see the creator in the creation, the source of all existence in and through what is bodied forth from that source." Her view allows for some distinction between God and the world, and thus parts company with Jantzen's more explicitly pantheistic approach. How helpful is the image of the body for an understanding of God, even in feminist terms? Bodies today become subjects that can be manipulated and intervened through medical practices and technology; consumerism pressurizes women to conceive of idealistic images of the body; cyberspace replaces the image of the body with a virtual world that is no longer subject to earthly constraints. All such cultural trends give bodiliness as such an ambiguity that then can overshadow any more positive advantages of such identification.

McFague, *The Body of God*, 133–34.

It is significant, perhaps, that McFague's subsequent book, *Life Abundant*, focuses much more on social and cultural issues of economics.[18] Like Grey, she is concerned with the culture of consumerism that dominates the Western world. However, she still holds to her earlier position of combining the agential and organic model through the panentheistic metaphor of the world as God's body.[19] Here she develops the idea of God's love as creator, liberator, and sustainer in a way that is concerned with creaturely flourishing as such, for her the glory of God expressed in the gift of life is concerned about the wellbeing of all creatures, in a way that challenges and seeks to transform current economic practice.

McFague argues for a prophetic and sacramental theology, but unlike Grey, McFague revisits Christology in such a way as to draw out these dimensions. Expanding upon the notion of "God with us" she suggests that an ecologically sensitive Christology centers on God as present with human beings as well as all other life forms. She suggests that it is the liberative and prophetic ministry of Jesus toward those who are oppressed that needs to be extended to all creatures, including the natural world. For her the sacramental dimension of Christology is both inclusive and embodied; the entire creation is *imago Dei*, rather than just human beings. She also believes, correctly in my view, that sacramental Christology adds a vital ingredient for contemporary discussion, namely that of hope. Yet for her the resurrection is interpreted as symbolic of the triumph of life over death; Christ's resurrection is "emblematic of the power of God on the side of life

18. McFague, *Life Abundant*.

19. Ibid., 138–41.

and its fulfillment."[20] The question in this case is whether such hope interpreted as akin to natural regrowth is sufficient to sustain us in the face of the terrible ecological tragedies. For her the resurrection expresses God's "yes" to all life in spite of suffering and pain. This is a different model of salvation compared with the traditional atonement images of substitution and sacrifice. For her sin is not just individual misdeeds, as in much traditional theology, rather it is the movement away from such flourishing, whether it be at individual or institutional levels.

An assumption in most ecofeminism is that patriarchy is linked essentially to dualism, which is a destructive way of conceiving relationships. However, Gillian McCulloch argues that patriarchal patterns of relationships are not inevitably and essentially linked with all forms of dualism.[21] She argues for a more sophisticated critique of dualism that allows for distinction in unity in a way that is actually more in tune with strands in feminist thinking that speak of difference.[22] In other words, a socio-cultural critique of patriarchy need not necessarily go hand in hand with a critique of theological dualism and associated pantheistic notions of God.

The drive toward activism may be partly related to anxiety among feminists that it is too reliant on white privileged women. One of the major criticisms of ecofeminism is that it has, like many branches of academic knowledge, been dominated by voices from the richer, Western nations of the world and led by predominantly white women. Ecofeminist writers have been self-consciously trying to address this problem by either living in

20. Ibid., 170.

21. McCulloch, *The Deconstruction of Dualism in Theology*.

22. Parsons, *The Ethics of Gender*.

or writing from different geographical and cultural locations. A good example of this genre that deals explicitly with Christian doctrine is a book edited by Grace Ji-Sun Kim and Hilda Koster on *Planetary Solidarity*.[23] One of the most striking aspects of this book are the stories told that are woven into the more theoretical accounts. Tools such as ethnographic analysis are brought to bear on women's environmental activism in the poorest regions of the world who literally risk their lives in order to protect and help those subject to environmental injustice. The systematic aspects of theology are treated through a feminist lens, as in earlier ecofeminism, but with a much more concerted effort to engage with the lives and livelihoods of particular cultural and social contexts and they use anthropological and other social scientific and even natural scientific tools. Politics and critical social theory have always been part of mainstream ecofeminism, but there is an internal debate about the extent to which certain aspects of the Christian tradition are rejected or retained. In other words there seems to be a fracture here opening within ecofeminist thought.

Many ecofeminist theologians are appreciative of Pope Francis's encyclical but bewail both the fact that he is silent when it comes to gender questions and that he displays a stereotypical view of women, including an idealist portrait of the Virgin Mary. He ignores the disproportionate impact of environmental harms on women in particular, which is a troubling oversight. While Pope Francis intends to be inclusive in his approach to care for the earth, feminists, at least, believe that they have not been heard. There may be good reasons why he has not chosen to address this problem, opening up, as ecofeminism

23. Kim and Koster, *Planetary Solidarity*.

necessarily does, challenging questions about how to interpret the tradition. More radical thinkers such as Heather Eaton, for example, push even harder against the Christian tradition and argue that Christology and its narrative of the resurrection and eschatology is no longer feasible for a truly grounded theology. For Eaton the idea of an afterlife enhances the dualisms that feminism has fought so long and hard to resist. Others, including Sallie McFague and Ivonne Gebara (I would count myself in here too), believe that Christianity shorn of faith in the resurrection, expansively and inclusively understood, loses its faith basis and therefore no longer provides the ground for that hope for renewal of the earth that is possible when perceived in the light of the resurrection.

REFLECTION

1. What kind of liberation theology do you think the Western world needs today?
2. Are there any examples of gender related oppression connected with environmental injustice in your region or state? How do you think these problems need to be addressed?
3. What theological advantages and drawbacks do you find in ecofeminist ideas such as that of McFague that the earth is God's body?
4. Is belief in the resurrection a hindrance or help for environmental activism?

4

POPE FRANCIS

Icon in the Anthropocene Era

This chapter will discuss the latest encyclical of Pope Francis, Laudato si',[1] *and concentrate in particular on its theological elements and its significance for ecotheology more generally.*

Pope Francis has, whether he wanted to or not, become a celebrity the world over. His celebrity status means that his message cannot be ignored, quite regardless of whether conservative or radical Catholics challenge some of his ideas. A proportion of more conservative Catholics find his approach offensive as it appeals to the grassroots and it challenges capitalist economic patterns of production and consumption. A further proportion of radical feminist thinkers object to his treatment of women, or rather, his gender silence. But his approach is nonetheless

1. This chapter will use the primary source *Laudato si'* extensively and paragraphs reference the encyclical. Pope Francis, *Laudato si'*, also available online from the Vatican website.

a contextual theology of the people that takes its inspiration from an Argentinian variant of Latin American liberation theology. While he lacks the philosophical sophistication of his predecessor Pope Benedict XVI and upsets what had been previously held to be the *status quo* for Vatican policy, his passion and broad insight across a range of subject areas make up for any shortcomings. Those who still resist his message about the environment brush off his attention to ecological matters as the failure of his advisors. The timing of the encyclical *Laudato si'* was critical, coming just before the international Conference of Parties Paris agreement (COP 21) on climate change. However fragile that agreement now looks at the start of a new Presidential era in the United States, Pope Francis has achieved something that no other popes before or since have achieved, namely, an openness toward the constructive input of theology in public debate on science and the environment.

In spite of some resistance just before its release, with the publication of *Laudato si'* Pope Francis finally put ecotheology firmly on the map of Christian theology and official Catholic discourse. Pope Francis intended this document to be read widely and discussed by those who are not necessarily Roman Catholic or even Christian. His appeal begins, like most other ecotheologies, with an observation of what is happening to the world around him, what he terms "our common home." His particular style of ministry gives his message further authenticity; here is a pope who refuses to be shielded by batteries of bodyguards, who takes homeless people off the street to share a meal with him on his birthday, who refuses the grandeur of his office, and who opts to live in shared accommodation. When he speaks, therefore, about the need for a simpler lifestyle few doubt his integrity.

BOX 1: FRANCIS OF ASSISI AND POPE FRANCIS

The theological and spiritual inspiration for Pope Francis's stance comes squarely from his namesake, Saint Francis of Assisi, the patron saint of ecologists. The name *Laudato si'* comes from the first line of the *Canticle of the Sun*, one of the best-loved songs in the history of Christianity, hence the first word "praise." Commenting on Saint Francis, Pope Francis says, "Just as happens when we fall in love with someone, whenever he would gaze at the sun, the moon or the smallest of animals, he burst into song, drawing all other creatures into his praise" (§11).

LAUDATO SI': THE MORAL DEMAND

There is urgency and verve to Pope Francis's writing so that he does not mince his words or try to wrap it up in a way that would make his message more palatable. He is aware that the media "at times . . . shield us from direct contact with the pain, the fears and the joys of others and the complexity of their personal experiences" (§47). He takes his readers to the root of the problems identified, both physical, such as turning the earth into a "pile of filth," and moral, such as a common indifference to the needs of the poorest and most oppressed peoples in the global community. Behind such indifference there is another kind of attachment that he believes bedevils humanity in a way that is becoming much more widespread in the global economy, and that is an attachment to technologies, social media, and other forms of interaction that allow us to avoid human contact and become distanced from the needs of others. Although he does not use the language of idolatry, this is implied.

In keeping with the liberation theme of the last chapter, Pope Francis's focus throughout is on both people and planet considered together, so protecting one does not make sense without the other. The first part of the encyclical lays out the ecological devastation now wrought on Earth that ecotheologians and environmentalists have been pointing out for nearly half a century. Pope Francis challenges the idea that human-induced climate change that is responsible for "most global warming" reflects a particular partisan view, since those severely impacted "have no other financial activities or resources which can enable them to adapt to climate change or to face natural disasters, and their access to social services and protection is very limited" (§25). Loss of livelihood leads to forced migrations. Lack of access to clean water by those in many parts of the world is a travesty of basic human rights. So:

> Our lack of response to these tragedies involving our brothers and sisters points to the loss of that sense of responsibility for our fellow men and women upon which all civil society is founded. (§25)

So what kind of changes does he recommend? Building resilience to climate change is not enough; concerted effort at mitigation is necessary as well. He is clear in recommending policies that (1) drastically reduce carbon dioxide emissions, (2) give greater access to renewable sources of energy, and (3) promote more widespread use of cleaner, less polluting technologies. Just as the world is losing its nerve about the possibility that collective human action on mitigation is possible, Pope Francis gives no such hesitation. He insists that current models of production and consumption therefore have to change, and change fast. But he is realistic about the present state of play, commenting that:

> With regard to climate change, the advances have been regrettably few. Reducing greenhouse gases requires honesty, courage and responsibility, above all on the part of those countries which are most powerful and pollute the most. (§169)

His cry is the cry of the prophet, calling out to all people to wake up to what we have done, instead of being smothered by indifference and apathy. So, when it comes to environmental challenges, "obstructionist attitudes, even on the part of believers, can range from denial of the problem to indifference, nonchalant resignation or blind confidence in technical solutions" (§14).

BOX 2: THE MORAL STATUS OF THE EARTH

Pope Francis's approach to the earth and its creatures represents a shift in favor of giving those creatures a higher status, comparatively speaking, though the notion that the earth has some sort of agency is also buried in some of the earlier encyclicals. He is certainly rather bolder in his resistance to unwarranted anthropocentrism as we move deeper into this encyclical compared with his predecessors (§115), but at the same time he wavers a bit in comparison with some of his earlier statements at the beginning of *Laudato si'*, where he uses the more instrumental scientific language of "ecosystemic services" to describe the contribution of other creatures to our common home. (§25)

Yet overall, his treatment of the scientific literature is balanced,[2] coming down firmly on the side of those who support climate change, with adherence to the consensual

2. Deane-Drummond, "*Laudato Si'* and the Natural Sciences."

view of the vast majority of scientists that human influence is primarily responsible for the sharp rise in greenhouse gases since the industrial revolution.

It is conservationists and ecologists, though, who will find the most power in what he has to say here, calling on them to do more research, find better ways of using energy, and protect life in all its diversity. Once we start to praise God for the created world around us, we find a different gloss on what we *see* around us, on our perception of the world. He presents a new vision that ecologists and conservation biologists have been sensitive to for years, but now understood through a theological lens. Humans are not alone in the world, but other creatures are all around us. He affirms each and every creature on the basis of their worth to one another and to God. So, "because all creatures are connected, each must be cherished with love and respect, for all of us as living creatures are dependent on one another" (§42). The goodness of creation is such that "the Church does not simply state that other creatures are completely subordinated to the good of human beings, as if they have no worth in themselves and can be treated as we wish" (§69).

His overall message that we and all other creatures are caught up in the common problem of our own making will strike chords with those who have been working in this area ever since Rachel Carson published her *Silent Spring* in the early 1960s. But will this be a watershed document in the same way that *Silent Spring* was for that generation? Will it really wake up those who are slumbering in their own worlds, too caught up with an obsession with consumerism to notice what is happening around them? The difference, of course, is that for Pope Francis an underlying faith in Christ and hope in the divine providence of God as Creator gives special reason for his firm belief that another world is possible. It is, I suggest, timely in that the last half-century has tried and repeatedly failed to find any way to move out of the trajectory of relentless progress at any cost that has locked in industrialist, capitalist societies bent on exploitative forms of profit for so many generations.

Lurking in the background is a critique not simply of particular rival political parties, in the US context parsed as Republican and Democrat, or in the UK Labor and Conservative, but a wider critique of how the acceptance of destructive cultural values undermines the very possibility of democracy as such. So, he claims:

> It is time to acknowledge that light-hearted superficiality has done us no good. When the foundations of social life are corroded, what ensues are battles over conflicting interests, new forms of violence and brutality, and obstacles to the growth of a genuine culture of care for the environment. (§229)

THE THEOLOGICAL VISION OF *LAUDATO SI'*

And Pope Francis, in keeping with the patron saint of ecology, Francis of Assisi, is not content just to be a prophet of doom, but provides concrete suggestions about how to act, along with a specific theological vision of an alternative. This is a theology informed by liberation motifs, even though he is comfortable using the language of "development," a word resisted by most liberation theologians on the basis that it presupposes a particular model of growth. It is clear that when Francis uses words such as "development" he does not mean the standard development model that apes the Western model of progress, but something very different, and much more akin to the liberationist perspective.

BOX 3: POPE FRANCIS AS LIBERATION THEOLOGIAN

Pope Francis's view emerges from the Christian option for the poorest members of the community, and is inspired by the example of Christ who came to serve the poor, even if social science is included somewhat sparingly in his treatment of economic or other social and political issues. Media portrayal of the Pope as dabbling inappropriately in politics ignores over a hundred years of Catholic social thought, beginning with *Rerum Novarum* (*Of Revolutionary Change*) penned by Pope Leo XIII in May 1891. In that encyclical Leo XIII critiqued both radical socialism and unrestrained capitalism, a theme common to *Laudato si'*. The point is that Catholic social thought seeks to inculcate a different way of thinking about the economy and politics, one that concentrates on the common good, understood as the good for all and the good for each. Pope Francis's diagnosis is that now the creaturely basis of that good needs to come into our analysis, for the simple reason that not only does all human life depend on the health of the earth, but those marginalized in human societies are also the ones facing the brunt of the impacts.

As well as a prophet, Pope Francis is also a priest. And he cares deeply for all people, including those caught up in relentless consumerism and addictions of all sorts. It is not people as such that he is against, but the way some humans fail to flourish in the way intended for the whole creation. And this means "respect must also be shown for the various cultural riches of different peoples, their art and poetry, their interior life and spirituality" (§63). In this he is open to those of other faith traditions in a respectful way. Behind such a concern he sees the need for a broken, modern world to be healed at the deepest level of human inter-relationships:

> If the present ecological crisis is one small sign of the ethical, cultural and spiritual crisis of modernity, we cannot presume to heal our relationship with nature and the environment without healing all fundamental human relationships. (§119)

Above all, he puts forward a theological vision of the whole that is consistent with traditional Catholic teaching, reinforcing strong and traditional Catholic concepts such as the dignity of all human beings, faith in God as Creator, hope in Christ who renews all things, and special devotion to Mary, who becomes not just mother of us all, but mother of the earth as well. She is a deep inspiration to those that share his Catholic faith, for "carried up into heaven, she is the Mother and Queen of all creation. In her glorified body, together with the Risen Christ, part of creation has reached the fullness of its beauty" (§241). [3]And it is this hope in the divinization of the whole earth that inspires practical action, for Mary as well as Jesus grieves for the "sufferings of the crucified poor and for the creatures laid waste by human power" (§241). So, he ends the encyclical with a reminder of the value of the family and the hope that, though mortal, humanity, along with all creatures, will one day share in the Sabbath of celebratory inclusiveness. Eternal life, the hope for which the human heart longs, is one of transfiguration and liberation

3. Feminists criticize this idealization of Mary, but it is one that at least partially resonates with the practical expression of Catholicism in some of the poorest communities of the world. For further discussion see Kim and Koster, eds., *Planetary Solidarity*.

from the bonds of sin and death (§243). It is joy that will accompany the struggle in this life, not the superficial joy of consumerist pleasures, but the deep joy of knowing that we will one day dwell in the presence of the eternal God, who is Lord of all life (§245).

If we rip out the theological threads from this encyclical we are left with a worn carpet that lacks the vibrancy that flows from a lifetime of prayer and contemplation. So it is appropriate that the afterword to this encyclical is itself a prayer, and a dedication to pray for the earth so that we come to recognize more fully how human lives are intricately linked with each other and that of other creatures. His prayer of praise to the Trinity—the Father, Son, and Spirit—undoubtedly came from his own pen, rather than that of his advisors. It is reminiscent, too, of the prayer of Saint Francis of Assisi, in speaking of the need for all humanity to become channels of love and peace in the world. Like other writers in ecotheology, Pope Francis insists, therefore, that healing of the planet requires first paying close attention to each other, by healing fragmented relationships. This is also the basis on which he urges the need for peace making in the face of gross injustices.

TOWARD ECOLOGICAL CONVERSION

How can we move from selfish attitudes to concern to build peace and ecologically stable communities? This, Pope Francis believes, requires a spiritual and moral conversion. He builds on the idea of ecological conversion that Pope (now Saint) John Paul II spoke about frequently, especially in association with the ecumenical Patriarchate Bartholomew I, about whom Pope Francis has warm words. It is remarkable, indeed, that each of the last three popes give space to the issue of ecology,[4] which then becomes a central platform of Pope Francis's ministry.[5]

As far as Pope Francis is concerned, and indeed Pope John Paul II before him, care for creation is an essential part of Christian faith and not simply an optional extra. So "all Christian communities have an important role to play in ecological education" (§214). This is unqualified, so he refers not just to some, but *all Christian communities*. The problem is too great to be relegated to a specialist interest of a few.

Leading from the idea of ecological conversion is a more constructive approach that Pope Francis articulates as integral ecology, which is also bound up with specific community building through dialogue between different groups. An example of that dialogue would be finding a ways of incorporating eco-justice insights into both human ecology and economics. So, economics, rather than treating ecological harms as "externalities" to a given course of human action, factors in ecological impacts from the start. This is practical and public theology writ large. John Paul II spoke of human ecology with an emphasis still on the *human*, where ecology formed a backcloth. Occasionally that slipped and he spoke movingly of the beauty of the natural world and how it inspires us to see God more clearly. On even less frequent occasions he viewed the natural world as personified. Pope Benedict XVI, on the other hand, developed the idea of an economy of gratuitousness, of gift, that implies an alternative to the current market model, but the explicit ecological implications were only hinted at. Any such haziness is gone in Pope Francis's encyclical. In *Laudato si'* we have

4. Deane-Drummond, "Joining the Dance."

5. Deane-Drummond, "Catholic Social Teaching and Ecology."

a clear, bold statement about humanity embedded in the natural world, an unambiguous requirement for Catholic Christians to care for the creatures of the earth as well as human families, and, like his predecessors, a priority toward those who are poor.

Pope Francis is also prepared to speak out against forms of human centeredness—varieties of anthropocentrism—that damage the natural world in a way that both of his predecessors were rather more hesitant to address. At the same time, this is not a radical break from Catholic social thought. Rather, it is a particular reading of the signs of the times through the lens of the mission and ministry that he believes he is divinely called to perform, namely, an imitation according to the life and work of St. Francis of Assisi.

What might that look like? It is fascinating that the themes of the *Canticle of the Sun* appear in this encyclical in reformulated form; for example, the *Canticle* begins with a message of praise, and praise is peppered throughout *Laudato si'*. Praise is given for creatures in particular, with references to the created world as a reflection of God the Trinity going back to a very ancient theological tradition from the dawn of Christianity, including writers such as Basil of Caesarea, Maximus the Confessor, and Thomas Aquinas.[6]

Can we still retrieve such high thoughts about the created world bearing an imprint of the Trinity while holding to the discoveries of modern evolutionary biology? Pope Francis clearly believes that we can, since he is prepared to accept the evolutionary account for creatures, even if he adds that human beings, "also possess

6. For an excellent discussion of the relevance of traditional Christian sources for environmental thought, see Schaeffer, *Theological Foundations for Environmental Ethics*.

a uniqueness which cannot be fully explained by the evolution of other open systems" (§81). But what does he mean? He is not saying that humans have appeared simply by divine fiat, but there is something about the way humans have evolved which is hard to pin down according to standard evolutionary theory. He is not controversial in this respect, since evolutionary anthropologists are still trying to work out what makes human beings distinct from other animals. Gone, however, in Pope Francis's account is an attitude of domineering superiority of human beings over other creatures. For him the specific and distinctive inventive, interpretive, and creative capacity of humans, along with our ability to create arguments, cannot just come down to "physics and biology."

The beatitudes say that the meek will inherit the earth, but what kind of inheritance might future generations expect? The *Canticle of the Sun* speaks of wind, air, and storms; change in climate appears early on in this encyclical. Saint Francis then moves to a discussion of sister water, "humble precious and pure," and a section on water then follows. Of course the sequence is not precise, but the point is that the themes in the *Canticle* are clearly a source of inspiration.

What other theological threads might we find if we turn over the tapestry and look at the way this encyclical is woven together? One central thread is that of love; a love rooted in the love found in marriage and the family, but extended to include others and the natural world around us. He is careful to be inclusive when speaking about humanity, so that he includes men and women, rather than just men. As mentioned in the last chapter, he could have done more to acknowledge that the burden of problems in the poorest nations of the world falls disproportionately to women.

Is Pope Francis too idealizing in his description of love for the created world? He speaks of the suffering of the natural world as a result of human interventions, but far less about suffering during the course of evolutionary or natural ecological history. I believe there is a reason for this. As a scientist he is well aware of the horrors of predation. Humans are not responsible for that, but we are responsible for speeding up the rate of extinctions, including extermination of peoples, cultures, and other creatures.

He also recognizes that some death is inevitable in order to live, but we should not take the death of any other creatures so that we might live lightly. Although he does not spell this out, the practical implications of such a view in terms of killing animals does depend on the criteria of restraint that he sets, whether that killing could be called necessary for human life or not.

For Pope Francis, it is clear that many of our habits of eating and consuming are not essential for life or health in view of the gross wastage of food in modern societies, unnecessary levels of meat consumption, and so on. He cites the Catechism which states that animals should not have to die needlessly. But our lifestyles do not mesh with this demand, and genetic manipulations of all kinds are not used appropriately. He could perhaps have gone further in his analysis of the social problems associated with transnational corporations and their exploitation of weak governance structures in developing outlets for genetically modified foods in poorer regions of the world.

Another important theological thread that Pope Francis touches on is humility. That comes through not just in his writing, but also in his willingness to draw on research that is outside the usual material of the encyclicals in engaging with scientific, ecological, and conservation literatures. Economics and politics are a common theme in Catholic social thought. What is much less common is consideration of the natural sciences. Pope Francis wants all of us to be more like that, to draw on the Catholic tradition but to be open to the insights of others rather than shoring up refuges of our own making, hiding behind either Scripture or tradition.

Instead, Pope Francis is convinced that tradition needs to become salt and light empowering a wavering world to act now before it is too late. Faith and hope are threads that run through this encyclical and insist on being heard. As he says, "all it takes is one good person to restore hope!" (§71). In his own person and action he has to some extent exemplified this model for the Catholic community and for the world at large. The Catholic Church has moved into the public sphere on environmental issues. The world is thirsty for a message of hope that comes from one who lives out that hope by living simply, in solidarity and in communion with creation.

Another thread that is perhaps harder to see is what I would term *glory*, but it is a glory often cast in the language of beauty. Even those who do not share a religious belief recognize natural beauty; though the paradox of the Christian life is that sometimes beauty appears even in what seems to us as aesthetically ugly. This is the message of the gospel, that suffering can also be productive suffering where love and mercy are shown to the afflicted. Pope Francis is confident that different natural species all contribute to the overall glory of God, so that their loss by extinction, whether caused by human or other means, is a diminishment of that glory. To put it simply, paraphrasing Psalm 104, all creatures give God glory just by existing.

Can we ever glimpse that glory of what Pope Francis refers to as sublime communion? Perhaps that glory is visible through another thread, namely, that of *joy*. Joy is not the same as happiness. We might feel a fleeting sense of happiness upon acquiring possessions of one sort or another. But the grip of consumerism is such that this kind of happiness is peddled as something else, namely a necessity rather than just a want. Contemporary societies have confused wants and needs. While many technologies have given us great advantages in all sorts of ways, including medicines and new inventions that make up our modern world, they can also be sources of inappropriate power, and possibly a kind of addictive force. We read that "technology tends to absorb everything into its ironclad logic," and we need liberation from this paradigm in order to become "more human, more social, more integral." Further, "compulsive consumerism is one example of how the techno-economic paradigm affects individuals" (§203). So "many problems of society are connected with today's self-centered culture of instant gratification" (§162). All of this leads to potential breakdowns in family and societal relationships.

A change along the lines that Pope Francis is suggesting is really hard to achieve. The pressure on young people today to accumulate gadgets is very strong. Some scientists say that the use of texting has rewired our brains. We literally become what we do. Getting back to deep appreciation of the natural world is tough going, and we have to make specific choices to address this, along with a deliberate restraint not to buy or consume just for its own sake. True joy comes from being liberated from the kinds of pressures consumer demands make on us, and from the freedom to experience the presence of God in each other and in the creatures with whom we share our home.

With those threads of love, humility, faith, hope, glory, and joy, in the spirit of praise we begin to find something else that Pope Francis also highlights in this encyclical, namely peace. Peace through dialogue is also given a name in Scripture that comes up toward the end of the encyclical, namely, Sabbath where there is possibility of *Shalom*. The Sabbath in the Hebrew Bible is not just a collapse of exhaustion, but a celebration of life, of the fullness and richness of life. The Sabbath rest gives us holy time, a time apart to be with God and each other. Why have North Americans in particular given up on the idea of a day of rest? Or even a rest from work? In the midst of the violence that comes to us day by day in the news media, something else needs to be in our hearts: a spirit of gratitude and thanksgiving, a spirit of praise. This is the legacy that the encyclical leaves us with, and one that should not be suppressed if we are to recover sufficient energy to deal with the complex environmental problems that threaten people and planet. Truly Pope Francis is an icon for the Anthropocene; his figure of humility is sorely needed in an epoch that has been named after us.

It is fitting, then, to end this chapter with the very final words in this encyclical as the last word, a cry from the heart: "O Lord, seize us with your power and your light, help us to protect all life, to prepare for a better future, for the coming of your kingdom of justice, peace, love and beauty: Praise be to you! Amen."

REFLECTION

1. What does integral ecology mean? Is it possible to achieve this in your own neighborhood? (See also Chapter 7.)
2. What kind of wisdom do you think Pope Francis displays in *Laudato si'*?
3. If you were given the task of writing an encyclical on ecology, how would you approach the task and what specific theological ingredients would you include?

5

DEEP INCARNATION

Christ in Ecological Perspective

> *This chapter will discuss how to understand who Christ is in the light of ecological concern and through the notion of deep incarnation, as discussed by key authors such as Niels Gregersen and Elizabeth Johnson.*

THE CHURCH, LIKE ANY social institution, has an inner committed core surrounded by the majority who are more on the fringes. Christmas and Easter celebrations remind us of that reality: suddenly a church becomes overflowing with newcomers, some of whom, perhaps, will decide to stay, but most of whom do not. What does Christmas *really* mean, once we strip away the tinsel and glitter? Does the coming of Christ have something significant to say not just about who God is, but about how to interpret that faith in an ecologically sensitive age? This chapter is really for those who are already committed Christians but want to know more about how to connect that faith with ecology. It is fairly obvious why belief in a Creator God will affirm the created world: if God has created it, then as believers in God we should love that creation as God loves it too. Creation is a gift. But the natural world as we know it also suffers and is far from perfect. Is there hope for that world that arises from Christian belief or not?

Pope Francis's encyclical covered in the previous chapter pays relatively little attention to Christology. Why? My own interpretation is that his more explicit Christian and Christological message is delayed in order to win over more readers who are very likely either neutral or perhaps hostile to Catholic teaching. This is particularly the case in the Western world that is secular, even if it is open to alternative forms of spirituality. He also wants to place his encyclical in line with other Catholic social thought that concentrates specifically on practices. His strategy was a deliberate choice, rather than a gap in his perspective. Further, his concluding paragraphs deal with cosmic liturgy and Eucharist, which implies that he is open, at least, to the much broader interpretations of who Christ is, traditions that, as we will discover in this chapter, also go back to the early church.

The incarnation—the idea that God has become human in the conception and birth of Jesus and yet still is divine in Jesus Christ—makes most sense to those on the Christian "inside." I do believe, though, that as long as sufficient care is taken, this belief can still be helpful in dialogue with those from other religious traditions. I will always remember my surprise at an international Muslim—Christian dialogue meeting on science and religion when a Muslim scholar was most interested in those very aspects of Christianity that made it different from his Islamic faith. There is not a hard and fast rule about what is a good choice of topic to bring into interreligious discussion. What is clear, however, is that Christian theology, if it is to be Christian, cannot ignore the incarnation.

This chapter will cover what has been dubbed not just incarnation, but *deep incarnation*, a term that was coined by philosophical theologian Niels Gregersen in order to discuss the evolutionary significance of who Christ is.[1] The history of the Christian church is built on acknowledgement of Christ's significance for human life; Christ is the one who, as savior, delivers humanity from evil, restoring a broken relationship with God. Classical faith in God as Creator, on the other hand, presupposes first the absolute difference between God and creation, even though God is the ultimate source of all that exists, for God creates *out of nothing*. Secondly, classical faith maintains that the created world is sustained in being through the ongoing presence of God as immanent in all that exists.

What happens, then, when God becomes physical material, enfleshed, taking on human "nature" in the historical person of Jesus Christ? Ecotheologians want to recover the idea that Christ's coming is not just significant in restoring broken sinful humanity in their relationships with God and each other, but also is significant in restoring humanity's broken and sinful relationship with the natural world. Further, extreme suffering (not just mortality) has existed in the natural order even prior to the appearance of humans. Clearly, that suffering is not directly the result of human activities. The movement is different, in so far as "sin," understood in terms of a broken relationship with God, if it exists at all in other

1. For a more detailed and academic discussion see Deane-Drummond, "The Wisdom of Fools?"; and Deane-Drummond, "Who on Earth is Jesus Christ?"

creatures, is not the same as that in humans who are capable of deliberate choices. Humans are the arch sinners, arguably the first to deliberately sin and turn away from God in their pride and arrogance, as the story of the fall of humanity testifies. But if we believe that "heaven" is for all creatures, which is reasonable, if not normally assumed in the classic tradition, then all creatures are also in need of healing from suffering. Hence, *redemption* is significant for all creation, even if *soteriology* (release from sin) in the strict sense applies to human beings. What arguments could be brought to bear in order to support this wider redemptive movement? One argument is that from an interpretation of the incarnation that is actually very ancient and characteristic of the early church.

There are two meanings of incarnation worth clarifying. "Strict sense" incarnation refers to the incarnation of God in the physical body (*sarx*) of Jesus Christ as a single human being. "Broader sense" incarnation is inclusive of other beings in the sense that Christ also shares the social and geo-biological conditions of the whole cosmos. In one sense all humanity shares this as well through the ancient idea of humanity as a *microcosm* of the *macrocosm*. In broad sense incarnation the significance of Christ as microcosm of the macrocosm takes special shape. Eco-theologians who affirm deep incarnation push the idea of incarnation beyond strict sense incarnation by reminding their readers of the wider significance of who Christ is for the whole cosmos.

BOX 1: WHAT IS DEEP INCARNATION?

Niels Gregersen was one of the first to use the term deep incarnation, and he applied it to the specific case of understanding Christology in evolutionary terms. He was also aware of the importance of this term for other practical situations of ecological importance, including climate change. For him, Christ entered into the "whole malleable matrix of materiality." Drawing specifically on the ancient Greek meaning of "flesh" or *sarx* as indicative not just of vulnerable bodies as in modern usage, but a much wider scope to include cosmic reality, Gregersen argues for the significance of the Word made "flesh" as encompassing the natural world from the very beginning of the cosmos right up to the present day. Gregersen's definition of deep incarnation is

> the view that God's own Logos (Wisdom and Word) was made flesh in Jesus the Christ in such a comprehensive manner that God, by assuming the particular life-story of Jesus the Jew from Nazareth, also conjoined the material conditions of creaturely existence ('all flesh'), shared and ennobled the fate of all biological life-forms ('grass' and 'lilies'), and experienced the pains of sensitive creatures ('sparrows' and 'foxes'). ("Christology")

See Gregersen, "The Cross of Christ"; Gregersen, "Christology"; Gregersen, *Incarnation*.

Those ecotheologians who ascribe to some version or other of deep incarnation draw heavily on biblical sources, especially the Prologue of John's Gospel, which sets the stage for the rest of the Gospel. Just as there are debates among biblical scholars as to the meaning of this text, so there are debates among ecotheologians as to the meaning of deep incarnation premised on particular biblical interpretations. Gregersen, for example, puts the most emphasis on the Greek Stoic background that is highlighted in the Copenhagen school of New Testament scholarship. According to this school *Logos*, translated as Word, is associated with the Greek term "in the beginning" (*en archē*), and thus reflects both the principle of the foundation of the universe and its continuity. This immediately brings a cosmic scope to Logos.

Early Christian writers, such as Tertullian, also drew on Greek philosophy for an interpretation of the meaning of the Logos.[2] Gregersen's interpretation of the Logos as divine information that is then echoed in the whole of the created order is heavily informed by Stoic interpretations, even though he recognizes that Christian faith that the Logos existed *before* world came into being parted company from Stoicism.

The Hebrew text related to flesh in the Johannine phrase, "The Logos became flesh" also extends out beyond the human community. The Hebrew idea of *kol-bashar*, all flesh, could mean human beings (e.g., Ps 65:3; 145:21), or all living creatures under the sun (e.g., Gen 6:17; 9:16–17; Job 34:14). So both ideas of "in the beginning" and "all flesh" seem to point to a much wider creaturely and cosmic significance of the coming of the Logos than just the human community.

At this point we need to ask: how might such a move encourage ecological responsibility? Does it make any difference at all if Christ's significance is understood more broadly? If Christ identifies with the whole suffering earth, including evolutionary and ecological aspects, then there is a shift to a sense of divine solidarity in suffering. But do the cosmic elements serve to take away from the historical grounded life of Jesus Christ and therefore tend toward abstract speculation, especially with the idea of a pre-material Logos? The challenge for ecotheology is to make Christ relevant for the whole cosmos, but not thereby to lose a sense of grounded, material reality. The

2. For a very interesting article on background to cosmology, see Walls, "Cosmos."

point is that if Christ is in some way identified with the earth, then Christians have an added reason to care for that earth, quite apart from belief in God as Creator. The alternative, to stress pantheism, is a route followed by some feminist scholars.

Feminist Roman Catholic theologian Elizabeth Johnson has provided an alternative to pantheism and Hellenistic Christology through her own interpretation of deep incarnation.[3] Johnson is, however, more explicitly concerned with prioritizing *concrete* ecological relationships and implications for ecological ethics. She emphasizes the Word made *flesh* to stress the transient, finite nature of the incarnation. Like many other writers on ecotheology, especially those informed by the work of Thomas Berry, Johnson is fascinated by the evolutionary origins of life and the interconnectedness that ensues; human beings and all other life forms are literally stardust. It is this rich sense of continuity and connectedness that for her makes the concept of deep incarnation particularly apt ecologically. Thus she claims that the flesh that Christ became is also materiality shared with the cosmos. But Johnson recognizes a problem with a cosmological reading: the last two hundred years of Christological scholarship has put the most emphasis on historical aspects of Jesus's life and ministry.

The problem now arises: how to affirm a wider interpretation of the significance of Christ, but also keep in line with mainstream biblical interpretations? Johnson's solution is through the idea of Jesus having a "deep ministry," which means attention to people and the natural world. Further, it is Jesus' *loving ministry* that undergirds compassion for the whole created cosmos. Johnson then

3. Johnson, "Jesus and the Cosmos."

elaborates the idea of "deep cross" following reflection on the Jesus's passion in its solidarity with the violent suffering, including the suffering and death of all creatures. The significance of Christ's resurrection as "deep resurrection" also extends beyond the human to include all life.

Johnson has corrected Stoic conceptions of deep incarnation through deep ministry, deep crucifixion, and deep resurrection. But are there other ways of reading the Johannine text?

THE WORD BECAME FLESH

Biblical exegetes have argued for some time that behind the Prologue of John there is an ancient cosmology that portrays a particular *Weltanschauung* or ideological or metaphysical framework.[4] The Hellenistic and possibly Stoic influence is not under dispute in historical criticism.[5] In the ancient world, cosmology refers not simply to an understanding of the geographical or physical features of the world, but represents deeper reflection on the significance of that world and humanity's place in it.[6] Both Gregersen and Johnson focus on *sarx* in the phrase "The Word made *sarx*," rather than on "The Word." Logos in the Prologue is associated with cosmological themes characteristic of Genesis 1, such as *in the beginning, creation, light, darkness* and so on. But it is *also* associated with historical accounts of ancient Israel, more characteristic of the specifically Hebrew emphasis on the action of God in history, such as the *tabernacle, glory,* and *enduring*

4. Painter, "Theology, Eschatology."
5. Anderson, *The Riddles.*
6. Brague, *The Wisdom of the World,* 4–25.

love.[7] In this way, in the Prologue of John's Gospel the coming of Christ is viewed in clear continuity with Israel's history, but it is now placed in a cosmological setting.

BOX 2: WORD AND DEED

In the Gospel as a whole, the influence of a Jewish emphasis on contingency in the human and natural world exists somewhat in tension with a more Hellenistic stress on universalism, but *all* aspects become woven into the Prologue of John. Hence, in the Jewish tradition Word is also associated with *deed*, so that Word always implies more than just abstract speculation. The "Word of the Lord" (*dābār YHYH; logos kyriou)* in Hebrew thought (such as Hosea 1:1 or Joel 1:1) had a particularly dynamic energy in conveying a double aspect of word and deed. In other places, the Word of the Lord was associated with life-giving (Deut 32:46–47); healing (Ps 107:20; Wis 16:12); illumination (Ps 119:105, 130); as well as being creative (Gen 1:1; Ps 33:6; Wis 1:1). In this sense the way that John uses the Logos terminology in the Prologue may be closer to the Hebrew *dābār* rather than the more abstract philosophical usages of the Greek *logos*.

This means, also, that the Word is associated with life from the beginning in its concrete expression. It is an "earthed" understanding of the Word. Contemporary philosophies of ecology are rather more inclined to stress contingent elements when compared with earlier ecological philosophies of even a hundred years ago that stressed the stability of ecological systems. It is the concrete, then, rather than the universal that is implied by *dābār*.

In John's Gospel, the "Logos" in some sense stands in for *Sophia*, or Wisdom (Hebrew *ḥokmah*), as Elizabeth Johnson has also acknowledged.[8] This background in Hebrew wisdom literature is important in helping to understand precisely what was meant by Logos when the Gospel was written. If Logos is really a "code" for *ḥokmah*

7. Rushton, "The Cosmology of John 1:1–14."

8. I am using the Hebrew term *ḥokmah* because I have found that students were sometimes confused by the difference between the Greek terms Sophia and Logos.

then this changes the meaning of Logos; Logos should be understood just as much in terms of *ḥokmah*. *Wisdom* is important as it brings a different slant on the way the universal and particular are united. *Wisdom* in the Hebrew Bible is fundamentally about right relationships between human beings, the natural world, and God,[9] while Logos reflects the reason or information in the world understood in terms of divine intent.[10] Both Wisdom and Jesus are sent by God into the world and "pitched their tent among us" (cf. Sir 24:8). John 1:14 interprets Jesus as one who lived among us, while Sophia delights in the human family (Prov 8:31) and seeks a place to abide in the created world (Sir 24:7). In the Prologue we find the passion narrative compressed into a few paragraphs, and it therefore anticipates what is to follow in the whole Gospel account. Wisdom is the cause of division and experiences rejection (Prov 1:20–33) like Jesus (John 1:11). Only some in a small community will accept Wisdom, and Jesus likewise draws together a community and shares a close relationship with his disciples. But the really crucial difference in relation to an understanding of the uniqueness of the incarnation is important. For while Hebrew Wisdom "appeared on earth and lived with humankind" (Bar 3:37–38), Jesus actually "became flesh." It is not theologically irresponsible, therefore, to suggest that such texts imply that in some sense Jesus is the incarnation of both the Word, *Logos* and Hebrew Wisdom, *ḥokmah*. Irenaeus understood Word and Wisdom as being like the two hands of God:[11] both are important in considering how God

9. Deane-Drummond, *Creation through Wisdom*; Habel, 'Where Can Wisdom Be Found?"

10. Maximus the Confessor developed the idea of *logoi* present in the created order as a reflection of the divine Logos. This echoes the difference between divine Wisdom and creaturely wisdom: both unite the universal with the particular, but the connotations of wisdom are more relational, while that of Logos is related to mind or Reason. For a discussion of the ecological significance of Maximus, see Bergmann, *Creation Set Free*.

11. Minns, *Irenaeus*, 51. His idea of the two hands of God may have come from Theophilus. In this interpretation Wisdom is associated with the feminine divine, the Holy Spirit, and Logos with Christ. This reading of the Trinity is problematic in so far as it

creates the world, and therefore both are important when considering the meaning of God becoming fully identified with the world in the incarnation. One of the difficulties in interpretation of this text is that the Hebraic connotations have been forgotten.

DEEP INCARNATION, DEATH, AND THEO-DRAMA

My own approach to Christology takes its cues from a reading of John's Gospel that puts its emphasis on deed, and a stress on *ḥokmah*, Wisdom, combined with the theological concept of *theo-drama*, drawing on the work of Hans Urs von Balthasar.[12] Von Balthasar's Christology is orientated toward the existential and experiential, but there are cosmic elements in his thought that are derived from the early church father, Maximus the Confessor. It therefore has significance for the relationship between Christ, ecology, and humanity as grounded in that ecology, though von Balthasar did not make these connections. Drama, when given an ecological reading, is more deeply grounded in scientific knowledge about ecology and evolution compared with von Balthasar. Many scientists, such as Evelyn Hutchinson,[13] have used drama as a way of describing the evolutionary unfolding of the natural order of creation. The incarnation is *deep* in the sense of being deeply embedded in the frail, suffering, and mortal

restricts femininity to the Holy Spirit and thereby may even have provided the occasion for the marginalization of Wisdom in subsequent history. See discussion in Deane-Drummond, *Creation through Wisdom*, 113–52.

12. I have developed this idea of a broader theo-dramatic reading in a number of places, most notably, Deane-Drummond, *Christ and Evolution*.

13. Hutchinson, *The Ecological Theater*.

history of the flesh of Christ. But as Johnson has been at pains to point out, it is a *particular and unique* history that is of importance in its universal and profound significance. The question that then comes to mind is: how and in what sense might that particular *history* relate to the wider history of other human beings, and beyond that, to the evolutionary history of all creatures in their ecological relationships?

Hans Urs von Balthasar recognized that theo-drama is not simply a narrative that happens, as it were, from outside human experience, but is embedded in mystical experiences from which human action flows. Drawing particularly on the insights of Ignatius of Loyola, he claimed that God is in all things, but at the same time the challenge or standard of Christ is one that relativizes human achievements, spurring the believer on to greater obedience to God's call. It is the capacity for human wonder in the face of the natural world that is a prerequisite to understanding or appreciating the significance of deep incarnation.

Hans Urs von Balthasar is of special importance because of his focus on the death of Christ as a crucial aspect of the dramatic action of God in history. Drama shows up the specific action of God in contingent events. Drama differs from grand narrative in that while it includes narrative elements, it focuses attention down onto the specific, contingent elements that are important. Narrative, in as much as it morphs into a grand narrative, carries a sense of inevitability and fatalism.[14] The category of drama does not take out all narrative elements, but becomes wary of false objectivity, the idea that it is possible to "stand outside" the story in which we, as human beings,

14. Deane-Drummond, "Beyond Humanity's End."

participate. Drama also includes a more contemplative element noted above, what one might term "lyric," which is a more mystical and existential way of interpreting events. Theo-drama is situated somewhere between narrative and lyric, and tries to avoid the dangers associated with both. Importantly, therefore, it mediates between what philosophers would term an ontological and a historical approach to Christology.

BOX 3: THE SIGNIFICANCE OF THE CROSS

The cross is of crucial importance for von Balthasar in the drama, so that "God's entire world drama hinges on this scene. This is the theo-drama into which the world and God have their ultimate input; here absolute freedom enters into created freedom, interacts with created freedom and acts as created freedom" (TD 4, 318). He understands such a drama as a revelation of the Trinity, rather than its actualization, so that such an action is a mirror of the immanent Trinity expressing itself in absolute self-surrender. In other words, for von Balthasar "it is the drama of the 'emptying' of the Father's heart, in the generation of the Son, that contains and surpasses all possible drama between God and a world" (TD 4, 327). In saying this, he seems to be trying to avoid the difficulty that could arise if the cross becomes somehow *necessary* as a way of describing Trinitarian relationships. It is therefore *self-giving love*, rather than the cross itself, that von Balthasar argues is at the heart of the immanent Trinity and at the heart of the incarnation. One way to understand deep incarnation, therefore, is not just incarnation into the mortality and fragility of human existence, but is *also* as a way of revealing what God is like: the deeply loving, self-giving and "emptying" God.

Von Balthasar, *Theo-drama*, 318.

Von Balthasar's idea of the primacy of God's love is also important as it gives an *ontological thread* in interpreting both Christology and the incarnation. From the beginning of creation, through to the incarnation and consummation, the Trinitarian movement is the dramatic movement of God's love and grace in the world. Creation, then, is not so much the backdrop against which human history is played out, but the *first act in the overall drama*, that eventually comes to expression in the incarnation of the Word (or Wisdom) made Flesh. In order to stay faithful to the heart of the incarnation as the Word/Wisdom made *flesh*, it is worth thinking more about the reality of death and mortality as well as interdependence between creatures.

Environmental ethicist Holmes Rolston III has used the language of "cruciform" to connect the suffering in the natural world with that of Christ.[15] This language would seem to fit with the language of deep incarnation in its broad sense, as developed by Gregersen and the "Christic" vision of authors like McFague and Johnson. But what is preferable, the language of theo-drama or "cruciform" nature? I prefer theo-drama for the following reasons:

1. Theo-drama looks beyond the event of the cross to the resurrection in a way that "cruciform" does not.
2. "Cruciform" marks out suffering and death as a necessary part of overall evolutionary and ecological processes. The way that I am interpreting the cross implies not so much the *necessity* of suffering, but its *inevitability*. This shift from necessity to inevitability is not about letting God off the hook by rendering the evolutionary process amoral, or refusing to face evolutionary accounts as currently understood. Rather, it is opening up to the possibility that the

15. Holmes Rolston III has referred to this concept on a number of occasions, but for an early example, see Rolston, *Genes, Genesis and God*, 306ff.

created world can, in theory at least, be created anew in a way that is not weighed down by that suffering. So, the vision of a new heaven and new earth is one that is a *renewed* earth, not totally new from scratch.

3. "Cruciform" fits more easily with the idea of evolutionary history as a grand narrative.

A *theological* necessity in relation to how God acts is just as inflexible as a natural necessity proposed by Rolston and Johnson. Both approaches, I suggest, weaken a sense of human responsibility for sin and therefore responsibility for the way humans treat the natural world.

BOX 4: HOLY SATURDAY AND ECOLOGY

Von Balthasar's theo-drama also includes a novel reflection on Holy Saturday. Christ's entry into Hades is an entry into the world of the human dead. As he enters Hades, Christ is sharing in the existential, human fear of death, and expanding von Balthasar's brief, we might include that fear and anxiety associated with ecological devastation and climate collapse. It is also a sharing in solidarity with those humans who are dying, perhaps as a result of ecological or climate devastation, or through forced migrations, and who, uniquely among other animals, share a profound fear of what lies beyond death. It is even a sharing in the knowledge that human beings are collectively bringing about the death of the planet. The difficulty, of course, is how far does von Balthasar's speculations about Christ confronting absolute sin in Hell represent an unfortunate type of *dis*incarnation, a removal from the Word made *flesh?*

I suggest that God's love in the face of death is much better expressed through envisaging God as acting *through improvisation*, even in solidarity with the world of the dead. The humanity of Christ is understood as a genuinely incarnate humanity grounded in the life of other creatures. But the continuance of human life beyond death is important inasmuch as it shows the capacity for humans to engage in an existential sharing with creatures beyond the grave. Christian religious experience is full of accounts of communion with the saints, visions, and so on as hints at the salvation history that is to come. The tradition has been slow, however, to acknowledge what this might mean for the earth as such, and not just humanity.

In this chapter I am making the claim, therefore, that deep incarnation should be understood not so much as the *spatial* descent of God into creation, or even the ontological *extension* of Christ into creation, but most profoundly as the *transformative and dramatic movement* of God in Christ who takes center stage in the theo-drama. Such a transformative movement is accompanied by the active presence of the Holy Spirit, so deep incarnation can be envisaged as an aspect of *pneumatology* as much as Christology, and points toward an eschatological vision of glory. Here we find pneumatology in the space between creation and re-creation, in the creation as it is now and the promised eschatological hope where God will be all in all. One of the reasons why this is important for practical ecotheology is that it gives humanity a place and active role in that drama alongside that of Christ who has trailblazed the path that Christians are called to follow. That path is one of suffering in solidarity in *imitatio Christi* with the poorest communities on earth, and the suffering of the dying, disfigured creatures whose extinctions are littered all around us. This leads to the topic of the next chapter, namely, a new anthropology for the earth.

REFLECTION

1. What does deep incarnation mean? Are you convinced by its importance for ecotheology or not?
2. Why is an understanding of Christology through the language of Wisdom significant for ecotheology?
3. What might be the first step in making imitation of Christ a reality in the ecological sphere?
4. How does deep incarnation treat creaturely suffering, death and extinction?

6

A NEW ANTHROPOLOGY FOR THE EARTH[1]

This chapter will cover new ways in which to conceive the significance of human beings made in the image of God in the light of current understandings of the entanglement between humans and other creatures and ecosystems.

My childhood memories are rich with associations with particular animals that became part of our household. My first pet, Saturn, a mischievous black cat with a splash of white on her chest; my ponies, the dun gelding Kelpie, the dark bay New Forest gelding, Diamond, and the bay Irish mare Shamrock; as well as my childhood black Labrador dog, Zena, and our current family dog, Dara who, at age 3, died tragically, most likely from a tick born blood disease, as this book goes to press. My relationships

1. This chapter develops some of the ideas published in Deane-Drummond, "Windows to the Divine Spirit." For a more academic discussion see Deane-Drummond, *The Wisdom of the Liminal.*

with them were just as intense as the rest of the family; indeed my sister regularly remarked that she preferred horses to people. That sense of close intimacy with other animals and the mark that they make on us as humans is also common across different cultures and geographical locations, not just as domesticates, but through human associations with wild creatures as well. Why is it that such associations have not on the whole been taken seriously by ecotheologians? Is it because ecology is more interested in the system as a whole and therefore tends to forget the particular? Have we forgotten our deep history of encounters with other animals? This chapter begins to interpret such an encounter theologically, as a way of sparking specific interest in taking other creatures seriously.

While the intersection between humans and their pets or even wild animals is something many people can connect to, widening the understanding of *creatures* so that all life forms are included is important even when thinking about the relationship between humans and other animals, since their lives and ours are embedded in a wider ecological niche. My students in the class of 2016 undertook a practical tree project. Each student picked a different tree on the leafy and tree filled campus at the University of Notre Dame. Each week, for at least seven weeks, away from their phones and other devices, they were asked to observe closely in silence what they saw, note the position of the sun, and interpret this in the light of what they had been learning in class that week in a portion of their learning journal called the tree project. Reading these reports at the end of the semester was incredible for me. Many students had never before spent time with either a tree or a living creature in this way, brought up in city environments away from encounters like this. For others it reminded them of family associations with specific domesticated animals or landscapes and the intimacy they gained with their feline or other furry friends. Most significant of all was a great sense of attachment that happened quite naturally over the course of a few months. When asked how they would feel if this tree was sawn down, many expressed anguish at this possibility; it was "their" tree and they wanted to keep it safe. The lesson sunk in as well: if they care this much about a single tree, what about the rest of the suffering created world? It opened up a compassion that they didn't know they had.

The specific focus for this chapter is humans, but humans now understood theologically through *between species* encounters, especially those between humans and other creatures, especially other animals, but also situated in a wider dynamic, living ecological niche. Ecotheologians have focused for years on how to think about the interrelation between different creatures and their natural habitat, but getting to the particular implications for how humanity and other creatures might themselves be transformed through that encounter is left in the background. Ecotheology has also for many years insisted that human image bearing needs to be understood in *relational* categories, rather than particular functional classifications or according to specific capabilities, such as reason. That is why understanding deep incarnation explicitly in terms of a rational term, Logos, needs to be complemented by reference to the more relational Hebrew term, *ḥokmah*.

ENCOUNTERING CREATURELY OTHERS

The symbiotic presence of living creatures in each and every living human body—the *microbiome*—are a hidden presence now made visible by biological science. There is intricate wisdom here, certainly, in the mutual dependence within each complex form of life, for such associations reverberate throughout the creaturely world as we know it. There are also likely to be many thousands of relations that have not yet been found, given that our own knowledge of different creaturely kinds is still very limited. Scientists estimate that only a tiny fraction of the different creatures in the world come under close scrutiny and study; many are going extinct without even being named or identified. The task that God gave Adam to name the creatures in the Garden of Eden is still unfinished, but what might dominion (Gen 1:28) mean for those who were living largely off the land on a vegetarian diet? What I would like to open up in this chapter is the presence of others on the journey of human becoming, others that exist in communion with human beings in surprising ways. I am going to do that by examining three case studies; on human interspecies relations with hyenas, elephants, and macaque monkeys, respectively.

CASE STUDY 1: HYENA–HUMAN RELATIONS

Human beings have been associated with hyenas for as long as their history permits. This might seem surprising, since we are used to thinking of hyenas as disgusting animals, associated with carrion and death, and with whom our only response is likely to be one of fear rather than delight. I am beginning by naming hyenas deliberately in the context of this chapter, for it is unfortunately all too easy to romanticize the natural world. It is important to be realistic about other creatures in their rich diversity:

some we will want to affiliate with, others less so. Yet, it is also those encounters with dangerous predators that have shaped who we have become. It is for this reason that deep evolutionary history is highly relevant to a discussion of ecological relations. For if we have little sense of where we have come from, we may not know clearly enough where we are going or even should go.

It is only recently that hyena distribution has become constricted to those areas of the world where we know them today, predominantly in the African subcontinent. Even as far back as 4.4 million years ago there was an association between hyenas and one of the earliest of subspecies or hominins in the *Homo* genus lineage, namely, the *Ardipithecines*. If hyenas had not been present many, many more hominin remains would have turned up; the hyenas crunched up the bone remains as part of their feeding habits. But half a million years later, we still find hominin *Australopithecus anamensis* associated with hyena remains. 3.6 million years ago further evidence for such association continues. These facts are not contested; there is no doubt that these remains show evidence of mutual interaction in the course of a complex social co-evolutionary history. The damaged bones of our earliest ancestors had previously been interpreted as evidence for a violent hominin, a "killer ape," persisting in the literature for three decades. Instead, close examination of the bones revealed the marks on them were not the result of activity by killer apes at all, but these unfortunate hominins had fallen prey to carnivores—with the evidence pointing to the most likely culprits being leopards and hyenas.

Homo ergaster, *Megantereon whitei* and *Pachycrocuta brevirostris* (giant hyenas almost as tall as the hominins themselves!) all coevolved. Half a million years ago the giant hyenas became extinct and were replaced by spotted hyenas, *Crocuta crocuta*. It is remarkable that the only African species persisting in Europe after the various glacial cycles were hyenas and humans, with hyenas finally disappearing from Europe and Asia a little before ten thousand years ago.[2] Why is this so crucial? I suggest it is crucial because our deep human history has not happened in isolation from other species, but in coevolutionary contexts, and with, from our modern perspective, some surprising others. So while we can use a microscope to look at the way symbiotic species have helped shape why we are alive and make life possible, the lens of deep history brings in another perspective on the dynamics of associations that we can all too easily miss.

Like the symbionts in our bodies, such a view challenges narrowly conceived perceptions of the human as being *most human* in isolation from and in sharp separation from other species. Rather, these companions on our way have helped to shape and reshape human communities, alongside a mutual human influence on those communities of other animals.

BOX 1: HYENA DRAMA IN HARAR, ETHIOPIA

Australian anthropologist Marcus Baynes-Rock has done some fascinating work on human-hyena relations in the Muslim town of Harar. He develops an argument for what he terms the "interspecies commons" between humans and hyenas by very close and detailed observations of their behavior. He describes in great detail a dramatic chain of events after a specific hyena in that town became poisoned. At first the dying hyena attracted the attention of the local people, who tried matches, smoke from burning rags, lime, milk, and other attempts to revive the hyena. Eventually a fellow hyena "picked up the hyena in her mouth and

2. Baynes-Rock, *Among The Bone Eaters*.

marched off into the darkness with him. She was followed by thirty-one other hyenas, growling and whooping, their manes and tails bristling." The story continues. "On the second night after the above incident, there was some unusual activity at the place outside Yusuf's house. One hyena was uttering a series of low groans while the other hyenas present were agitated. They were scratching at the ground and gathering around at various places, sniffing together." A few minutes later about six hyenas arrived at the normal gathering space of hyenas at the Argobberi gate and appeared in an aggressive relationship with a group of other hyenas.

Baynes-Rock, *Among the Bone Eaters*; Baynes-Rock, "Life and Death."

Marcus Baynes-Rock unravels the puzzles of the behavior he observes by considering the social and biological significance of the entangled human-hyena relations. It is important to note that hyenas living in the city of Harar are entirely dependent on human food sources. Lions, which used to prey on hyenas, have largely disappeared in this region. Hyenas scavenge on left over carcasses, as well as occasionally break into unguarded livestock pens. Humans also feed hyenas directly, and the locals exploit this practice as a tourist attraction. Yusuf's house, the place where the rest of the pack found the poisoned hyena as described in the BOX above, was one such feeding place. Baynes-Rock suggests that this practice makes the hyenas bolder in the company of people, even though the kind of food they are given is much the same as they would find in garbage dumps by their own scavenging. The Argobberi gate where the hyenas clashed has a human history, but also a hyena history, as it is the site of exchange between three hyena clans. The different hyena clans occupy different territories in different parts of the town, and enter it through different gates. Human feeding is confined to the Sofi clan, and Yusuf thought the dying hyena was from that clan; but Baynes-Rock's observations on the subsequent angry reactions of the hyenas from another clan made this explanation much less likely. The Aboker hyenas were angry as the dying member of their clan was taken off and presumably killed by the Sofi clan who routinely gathered at Yusuf's house.

Did the Aboker clan know that one of their members was missing? Baynes-Rock refuses to draw this conclusion except to say that they would have been aware of the commotion a few nights before which had involved one of their clan, since the feeding place is so close to the Aboker territory. Additionally the Argobberi gate is the place where disputes between hyena clans are resolved. Baynes-Rock suggests that the Argobberi gate "is a mutually constructed, historicized, politicized, meaningful place in the minds of both hyenas and humans who participate in the dramas which are enacted there."[3]

In the minds of the local people these spotted hyenas are *agents* just like humans, for, according to them, "hyenas hold meetings, make supplications for food, and communicate detailed messages to conspecifics and to humans who can understand hyena language."[4] They also believe that hyenas will punish other hyenas who attack livestock, and humans who poison or kill hyenas will, in turn, receive retribution. When the hyena was poisoned (perhaps by accident) the local people treated it like a person in the same situation, since waving a match under the nose is standard treatment in this region for those with an epileptic seizure. Lime-juice is also standard treatment in those cases where a girl has swallowed bleach, which

3. Baynes-Rock, "Life and Death," 221.

4. Ibid.

occasionally happens where a girl is forced to marry against her will. The milk was Baynes-Rock's own idea, and Yusuf followed his advice after his own attempts failed.

Hyenas are also thought to be spiritually powerful animals that are able to mediate messages from the local town saints and pass these messages onto people that can understand hyena language. The messages are thought to be quite specific, including information about the number of *jinn* in the town, *jinn* being unseen spirits that can possess their owners or cause mischief. Hyenas are thought to be able to catch and eat *jinn*. This also explains why the locals recited the Qur'an after the hyena was taken away by the hyena pack. Further, it explains the long-term persistence of hyenas: far from being a threat, they are thought to be able to protect people from negative spiritual forces.

The connection between religious practices and the hyenas understood as agents in a theological drama is really fascinating. It implies that other animals in some sense are viewed as mediating agents with the divine. It is this mediating influence that also colors Christian religious history as well, for lurking in the story of Adam and Eve we find the ubiquitous appearance of the snake.[5]

CASE STUDY 2: ETHNOELEPHANTOLOGY

Elephants also have a long history of association with humans, representing weapons of war, prestige, symbols of divinity, entertainment, icons of conservation, vehicles for labor, and functioning as companion animals. Elephants are capable of considerable learning, retaining and transmitting practical skills, and social information.

5. Deane-Drummond, "The Birth of Morality."

Perhaps most poignant of all, it is the individual personalities of elephants that are capable of influencing patterns of interaction according to particular social conventions in the elephant community. So they can recognize each other as individuals as well as appreciate the thoughts and feelings of other sentient beings. Based on his own experience with observing elephants in Chitwan National Park in Nepali, Piers Locke makes no hesitation in challenging a common presupposition in much of the scientific and popular literature that animals cannot make judgments based on knowledge of the thoughts or feelings of others, sometimes termed "theory of mind." He also insists that elephants can grieve for lost friends and relatives. A research program called *ethnoelephantology* puts emphasis on subjective agency in both humans and elephants, their coevolution, and a method of research that interweaves the biological with the cultural.[6]

BOX 2: WILD ELEPHANTS IN CHINA

The last remaining wild elephants in China also play a significant, albeit unequal role in the network of interrelationships that link up scientists, government officials, farmers, conservationists, and tourists, especially those that deal with elephant-human conflict, species survival, and animal welfare. Human decision making about how to act wisely in this particular context is significantly *also* dependent on how elephants themselves act, including more destructive habits such as colliding with cars or destroying houses. Locals claim that elephants' protected status has led to them changing their habits and exploiting their even greater freedom. So in this context human indigenous rights start to be asserted afresh as a further *counter reaction* to elephant destructive behavior.

Locke, "Explorations in Ethnoelephantology."

6. Locke, "Explorations in Ethnoelephantology."

BOX 3: MACAQUES AND ETHNOPRIMATOLOGY

Agustin Fuentes has conducted a fascinating study of the long-tailed macaques living in the temple forest complex at Padangtegal, Bali. Here the boundary between nature and culture is blurred in what he terms "naturecultural contact zones." This association between humans and macaques is not as simple as either strict competition or purely reciprocal association between humans and other primates. In the temple complexes resident monkey groups are participants in religious ritual practice. They are viewed in diverse ways by the Balinese, either as just part of the environment, as a nuisance, or as emissaries of spiritual-natural forces. For Balinese the macaques are actors, just like humans. Macaques living in the temple complexes anticipate when they are allowed to feed on the ritual offerings from the long experience of their mutual association. On a biological scale, human alteration of the landscape has shaped the population genetics of the macaques in particular ways, where gene flow is channeled down specific riverine corridors created by the Balinese agricultural system. The main point here is that there is an *interlacing* of human and macaque histories so that their environments are constructed on particular lines, shaped by particular decisions made in human societies.

Fuentes, "Naturalcultural Encounters in Bali."

HUMAN BECOMING: THEO-DRAMA AND THE SPIRIT

I would now like to weave in a theological interpretation. How might we envisage the particular role of humanity in such exchanges? Where does, in other words, our *particular* responsibility as human beings lie? What does it mean to bear the divine image?[7] My own preference, given the cases I have discussed, is that this makes the most sense

7. Deane-Drummond, *Re-Imaging the Divine Image.*

through the category of *performance*, or *drama*, for the dynamic movement through that drama also speaks of the active presence of God in creation. Encounters, be they of joy or suffering, are woven into a theo-drama that reminds us that God is not absent from God's creation, but present with and through it in the rich entanglements of human and creaturely history. In theological terms this movement reflects an eschatological orientation toward redemption, for theo-drama takes its bearings from the life, passion, death, and resurrection of Christ. Such a constructive approach is connected with the ecological idea of ecological niche, *oikos* or home, but stresses its history and forward dynamism. A *niche* in ecology is the living environment in which a particular species is found. Human niche construction that charts the changing ecological relations of the whole system through deep history is not detached from other animal kinds, but is in some sense *entangled with* them.

This renders the theo-dramatic task as *inclusive* rather than *exclusive*, a way of seeing the active presence of God in relation to humans and other creaturely kinds in terms of *dynamic performance*, incorporating insights from evolutionary science, while still being self-aware of different methodological and metaphysical presuppositions.

Yet the relationship between particular biological and theological modes of expression resonate with each other and reflect an understanding of nature *as* creation. Talk about God is in a limited sense always *analogous* to what might be found and discovered in the human or scientific sphere, as we are using culturally bound *human* language even when speaking of humans as being made in the image of God. I am not suggesting here some kind of reduction to a naturalistic view, where the only compass for theological knowing emerges directly from scientific knowing. Nor am I suggesting that the analogy between what is discovered in evolutionary anthropology or biology is thin, such that the biological realities that are encountered are irrelevant or not substantive for theology and so can be ignored. Rather, the kind of knowledge that theology brings is not just discovered in the world of nature that is then interpreted as a created world, but is also *revelatory* knowledge, one that takes religious experience seriously, so orientated on faith in a loving Creator and Redeemer of the world. There are windows to that presence of the divine in the natural world, but such insights are also tested and wrestled with through the lens of Scripture and Christian tradition. I am not, therefore, prepared to assume that *everything* that is discoverable in the natural world is a sign of God's presence as Spirit. Rather, humanity has been given creaturely wisdom that helps discern when God's presence is encountered or not. The specific encounters of humans with other creatures are, at least in many cases, potential sites for recognizing that presence in so far as there is an acknowledgement of mutual interaction in the overall theo-drama.

REIMAGING THE DIVINE IMAGE

A Christian believer, or perhaps those from other religious traditions, especially indigenous traditions, will perceive more existential and religious significance in such encounters between humans and other creatures, but this imaginative religious capacity is itself a corrective to those dogmas that arise when science is falsely assumed to be value free. Sciences, including ecological science and evolutionary anthropology, no less than philosophy, are therefore like handmaids in the theological birthing of insights arising from paying close attention to the created world. A Christian will encounter the natural world with the background belief that God as Trinity is both creator and redeemer, and therefore will already be open to some extent to the possibility for a form of nature mysticism. That divine presence is received by those who are, through a kind of poetic sensitivity, attuned to the pathos of both beauty and horror in the natural world, so, like Gerard Manley Hopkins, ready to consider God in all things. The new creation that is dependent on the active presence of the Spirit is not, therefore, presupposing a total destruction of the existing order but its renewal. This also fits with the intent of the book of Revelation, where the new heaven and the new earth is not so much a replacement as a *renewed* heaven and earth.

Human beings in the image of God, by mediating between creaturely and divine, occupy a middle ground in historical terms as much as in spatial terms, as represented through ideas such as the microcosm. Image bearing is about human performance in relation to others, but, as divine image, humans bear a special responsibility to act for the common good as perceived according to divine Wisdom. The common good is not narrowly prescribed in the human community, but opens up in dynamic relation with other creatures. Humans are richly embedded in a life with others. Ethics, in other words, needs to take account of the deep creaturely reality from which we have come, only fragments of which arise through *paying attention* to that world by close and detailed observations and experiential encounters. It is to these ethical aspects that I turn to in the chapter that follows.

REFLECTION

1. What encounters with particular creatures or landscapes are part of your experience?
2. Have we, in the Western world, lost a sense of interdependence on other creatures?
3. What difference might it make theologically if other creatures share in a theo-drama with humans on planet earth?

7

CHRISTIAN ECOLOGICAL ETHICS

This chapter will discuss the specific contribution theology might make to a discussion of ecological ethics, including the economy, environmental sin, and ecological justice.

ONE REMARKABLE DEVELOPMENT IN ecological ethics has been the joining up of justice for the earth with that of oppressed peoples.

BOX 1: DAKOTA PROTESTS[1]

A proposed Dakota pipeline that began construction in 2016 is proposed to run from the North East corner of

1. I am grateful to Michael Schuck, Loyola University Chicago, who circulated slides of his visit to the Dakota Pipeline site, and the material in this box is indebted to the information he provided on December 16, 2016, personal communication. See also Christiana Peppard, "Laudato Si' and Standing Rock" and for a broader discussion of water justice issues, Peppard, *Just Water*.

North Dakota to southern Illinois, where it connects with pipelines to the Gulf of Mexico. The original route of the pipeline went through a predominantly white community. The new route goes through unceded Sioux territory under the 1851 Treaty of Ft. Laramie. On September 3rd archeological sites were bulldozed, leading to a cluster of camps on the border with Standing Rock reservation. The Oceti Sakowin Camp, along with the Sacred Stones Camp and 1851 Treaty Camp, are a center of protest. The pipeline work on top of Turtle Hill is visible from Oceti camp. The protests are not just symbolic, they are about the basic right to access to clean water that has come into conflict with the desire for fossil fuel access by a privileged elite.

The solidarity statement of the Bemidji State University in Bemidji, Minnesota reads:

"I oppose the construction of the Dakota Access Pipeline on the native lands and through the Missouri River because it poses an imminent threat to vital resources and violates the basic human rights of First Nation Peoples. I acknowledge the history of violence and oppression against indigenous peoples in our country and recognize the actions at Standing Rock as a continuation of this legacy of abuse."

So far I have covered theological arguments for integrating ecological ideas into theology; but what might that mean in practice and in the light of the specific issues facing real communities? Often the line between theology and ethics is blurred in ecotheologies. In this chapter we will delve into some of the more explicit ethical discussion of topics, and their philosophical basis. Christian moral theologians who venture into this territory cannot afford to be ignorant of either the scientific accounts of relevant trends, such as loss of biodiversity or climate change, or philosophical arguments for and against specific practices.

Environmental justice raises issues that are of social, legal, political, and religious relevance. Ecological ethics includes such concerns, but pushes further than this. It considers, therefore, not just the plight of those most vulnerable to environmental harms, but also wider ethical issues of ecological degradation, loss of biodiversity, ethics of food production, and so forth. Justice issues have always been a concern for moral theologians. Widening that brief so that the recipients of justice-making are inclusive of creatures and the land beyond the human community is part of the agenda of ecological ethics.

BOX 2: GLOBAL ENVIRONMENTAL JUSTICE

Environmental justice (or more properly, perhaps, *injustice*, though this term is not normally used) is the disproportionate impact of environmental harms on vulnerable populations. There are a vast number of examples of environmental injustice and these examples exist both nationally in the USA and internationally. In New Mexico, for example, the Navajo nation, living in the Carrizo mountain area, were heavily exploited through uranium mining projects in the region that was first opened up for its rich vanadium resources. According to their own religious traditions uranium should be left in the ground and not extracted. Workers were not warned about the hazards of uranium extraction, they were paid poor wages, and there was a lack of proper safety procedures in the mines. After a long struggle, the Radiation Exposure and Compensation Act (1990) was eventually put in place to protect workers in such circumstances. However, the US government imposed strict certification guidelines that were inappropriate for the context of the Navajo people.

Similar stories proliferate the world over and ongoing reports can often be accessed through fact sheets from non-government organizations such as the Sierra Club, available at http://www.sierraclub.org/environmental-justice. For a comprehensive collection of testimonials, stories and activist statements see Ammons and Roy, *Sharing the Earth*.

WHAT ARE APPROPRIATE ETHICAL CONCEPTS?

Just as there are many and various ecotheologies, so there are different possible approaches to environmental ethical decision-making. The question of environmental justice, raised in the case study above, is one that is popular among Catholic scholars, since it represents an expansion of a long history of Catholic social teaching on justice. Pope Francis's notion of integral ecology, for example, finds a way of incorporating insights on ecological ethics, while retaining a strong sense that the Church needs to insist on priority for the most vulnerable and those living in poverty, the "option for the poor." Difficult ethical analysis comes to the surface in specific cases, such as the restoration of an ecosystem after environmental degradation, or where there is a clash between the economic needs of a community and bio-conservation ethics. Here local history of the site is important, but so is the consensus of the community. Possible tensions also immediately arise when local needs clash with what might work at a global level. Rather than rehearse all the different secular arguments for ecological ethics, we will concentrate attention on the ways in which theological analysis might make a difference to these discussions.

The Franciscan tradition of contemplation on the Creator's gift of the created world has, as Pope Francis has shown us, much wider appeal than the Catholic tradition itself. Acknowledging all creatures as gifts to one another, rather than as instrumental objects for human use, shifts ethical discourse away from management towards appreciation and care. The art of paying attention to the natural world is not, of course, restricted to Christianity or to any one tradition within it, but finding ways within one's own tradition to make that attention specific is vitally important as a first step in an adequate ethical response. Contemplation arising out of the Franciscan tradition is inclusive rather than exclusive, in that it seeks to include other creatures in prayerful communion with human beings. It was not the specific ethical concern for creatures that first moved Francis of Assisi, but rather the power of considering Christ's wider significance for the earth in light of the incarnation and, following that, an appreciation of the praise of all creatures.[2] It is much harder to harm those we pray for, and the natural world and its creatures are no exception to this. It is through that contemplation, too, that the seeds of genuine love grow stronger. We may also experience the emotion of wonder, well beloved of the earliest environmentalists such as Rachel Carson.[3] But the ability to wonder from a theological perspective takes us on a journey that is transformative in a religious sense, for it is wonder disciplined through a particular lens, namely, the lens of the passion narrative of Jesus Christ. This is also attuned to the way Franciscan spirituality approaches the issue.

A second aspect of specifically theological environmental ethics now comes into view, namely one that is inspired by Christ as an example of righteousness and links closely with justice. What might that justice look like through a Christian lens? Certainly it needs to include concern for the most impoverished members of the human community. While local needs are important, the global interconnection of environmental problems points to the necessity of a global framework for justice-making alongside local democratic decision-making.

2. Nothwehr, *Franciscan Theology of the Environment*.

3. Carson, *Silent Spring*.

Prominent secular social theorists on justice include philosophers and economists such as John Rawls, Martha Nussbaum, and Amartya Sen.[4] Of these, only Nussbaum is prepared to name what the good might look like in terms of global decision-making, but she has yet to develop these ideas fully for environmental ethics.[5] The link between ecology and development in terms of a more positive shape for justice-making is set forth in Roman Catholic Social Teaching (CST) through development of the idea of *human ecology* and is subsequently elaborated in Pope Francis's *integral ecology*. In order to understand what these terms mean, it is necessary to trace their emergence in CST.

The Roman Catholic Church has a global reach and is arguably one of the most influential Christian denominations on a global scale. But where and why did ecological issues come onto the agenda in CST? Many ecotheologians have ignored such social teaching entirely and have assumed that it is problematically tainted by a supposedly anthropocentric bias. I believe that it is far, far more complicated than this, in that the retrieval of CST lends itself to high impact, even if much of its emphasis is more human centered than ecotheologians would normally accept. There are particular reasons why Pope John Paul II became concerned with ecological issues, and an important one was the link between ecology and development that he perceived perhaps sooner than many others.[6] This allowed him to develop his particular

4. Rawls, *A Theory of Justice*; Sen, *The Idea of Justice*; Nussbaum, *Frontiers of Justice*.

5. I have explored what this might look like in Deane-Drummond, "Deep Incarnation."

6. For further comment see Deane-Drummond, "Joining in the Dance."

interpretation of *human ecology*. In commenting on the value of preserving the natural habitat of other species, he remarks that "too little effort is made to *safeguard the moral conditions for an authentic human ecology*."[7] He is reinforcing one of the traditional aspects of Catholic social teaching, namely, that there is a deep *ontological basis* for moral law that is rooted in the doctrine of creation. Further, he suggests that it is the *violation* of this law that is the most fundamental cause of the ecological crisis. *Integral ecology* builds on the social concept of human ecology, but is more explicit in its inclusion of ecological and environmental concerns alongside an explicit call for reform of social structures, an ecological economics. Part of the difficulty, of course, is that it is not always easy to reach decisions that allow for an integral ecological approach; but that, at least, is the vision set forth.

Pope Benedict XVI, like Pope John Paul II, finds particular valence in linking ecological harm with a threat to world peace. In his World Day of Peace message of 2007, he drew a close parallel between ecological flourishing and human flourishing, setting forth a vision of what justice requires. It implies not just environmental justice, that is, concern with the disproportional negative environmental impacts on the poorest of the poor, but ecological justice as well, that is, concern for the wellbeing of other creatures.

It is worth stressing that a commitment to affirming the natural world and ecological practice is not just confined to Pope Francis as we covered in chapter 4. If anything, Pope John Paul II came close to a form of nature mysticism in some of his statements:

7. Pope John Paul II, *Centesimus Annus* §38; italics original.

> So, in beholding the glory of the Trinity in creation, man must contemplate, sing and rediscover wonder . . . Nature thus becomes a gospel which speaks to us of God . . . this capacity for contemplation and knowledge, this discovery of a transcendent presence in created things must lead us also to rediscover our kinship with the earth, to which we have been linked since our own creation (cf. Gen 2:7). This is precisely the goal which the Old Testament wished for the Hebrew Jubilee, when the land was at rest and man ate what the fields spontaneously gave him (cf. Lev 25:11–12). If nature is not violated and degraded, it once again becomes man's sister.[8]

Pope Benedict XVI understands the natural world as an expression of God's "design of love and truth" and the natural basis on which human life depends, given as a gift of God to humanity. Pope Benedict XVI is, however, much more explicit in spelling out the specific ethical dangers in a turn to nature, the danger of *pantheism*, or total identification of God with the created world, as well as arguing against the technological domination already noted by Pope John Paul II. Pope Benedict XVI is known for his sharply critical approach toward ethical relativism in the Western world, that is, the view that all ethical values should be welcomed and accepted as equally good. He also objects to forms of scientific naturalism or the belief that there is nothing beyond "nature," or the transcendent. For Pope Benedict XVI the dangerous presupposition of natural science is the denial of any sense of purpose in the universe, so "nature, including the human being, is viewed as the result of mere chance or evolutionary

8. John Paul II, "General Audience," §5.

determinism."[9] Rather, creation "is a wondrous work of the Creator containing a 'grammar' which sets forth ends and criteria for its wise use, not its reckless exploitation."[10] Understanding the natural world as the work of the Creator then promotes its proper treatment, though Pope Benedict XVI still puts an emphasis on the "use" of the created world.

BOX 3: ECOLOGICAL SIN

Another feature of theological discussions of environmental ethics that is distinctive compared with secular approaches is the idea of environmental degradation as a *sin*. Ernst Conradie believes that Protestants have not paid sufficient attention to sin in developing ecotheology and the potential for sin language for social diagnostics, though traditional approaches have acknowledged the importance of structural sin and repentance in more general terms for some time. The ecumenical patriarchate His Holiness Bartholomew I insists on *metanoia*, a turning away from practices that harm the natural environment:

> What is required is an act of repentance on our part and a renewed attempt to view ourselves, one another, and the world around us within the perspective of the divine design for creation. The problem is not simply economic and technological; it is moral and spiritual. A solution at the economic and technological level can be found only if we undergo, in the most radical way, an *inner change of heart*, which can lead to a change in lifestyle and of unsustainable patterns of consumption and production. A genuine conversion in Christ will enable us to change the way we think and act.

John Paull II and Bartholomew I, *Common Declaration*.

9. Benedict XVI, *Caritas in Veritate* §48.
10. Ibid., §48.

It is the moral imperative to act that is perhaps one of the most distinctive aspects of Christian theological approaches to environmental ethics. For Pope Francis, like Pope John Paul II and Francis of Assisi who inspires his vision, from the very beginning of his papal ministry in 2013, the root of this call is his commitment to Christ:

> [W]e also see the core of the Christian vocation, which is Christ! Let us protect Christ in our lives so that we can protect others, so that we can protect creation! The vocation of being a "protector," however, is not just something involving us Christians alone; it also has a prior dimension which is simply human, involving everyone. It means protecting all creation, the beauty of the created world, as the book of Genesis tells us and as Saint Francis of Assisi showed us . . . Be protectors of God's gifts.[11]

In naming environmental degradation as sin, we first have to recognize our guilt, and Ernst Conradie, for example, writing from a Reformed perspective, speaks eloquently about the specific need to recognize our guilt in relation to climate impacts.[12] Recognition is not always easy, since how we act day by day leads to imperceptible changes that then impact on our climate. I have therefore suggested

11. Pope Francis, *Mass*.

12. Conradie, "Confessing Guilt." More recently Conradie has initiated a project titled *Redeeming Sin: Harmatology, Ecology and Social Analysis/Diagnostics*, in which he argues that the category of sin is useful for confronting deformations in social structures that are integral to complex ecological problems such as climate change. Much the same has been said by liberation theologians who name structural sin as a key social category. The difference is that Conradie employs clear categories from the Reformed tradition on sinfulness and God's holiness and is more explicit in naming the relationship between that sin and ecological harms.

the need for a new term called "anthropogenic evil," or more explicitly sin, which recognizes the anthropogenic element in climate impacts, following the scientific terminology of anthropogenic impacts used by the IPCC.[13]

Climate change is a good example of the kind of ethical dilemma that is extremely complex and requires a range of perspectives if we are to have any hope of arriving at an adequate response. Christian reflection on hope, along with other important virtues such as faith, charity, humility, justice, temperance, and prudence marks a distinctive approach to developing environmental virtues compared with secular alternatives, and there are biblical imperatives for developing such virtues.[14] Behind such a hope is faith in God's providential care, but such care is not to be divorced from taking human responsibility for how we act. My own preference is to use practical wisdom or prudence as a way of discerning how we might decide what it means to act justly, to love sincerely, or to express temperance while allowing for generosity.[15] But courage in the face of adversity, or fortitude, is also going to become increasingly relevant as we face the need not just for attempting to stave off climate change, but also for adaptation to its accelerating impacts.

MOVING JUSTICE BEYOND THE HUMAN SPHERE

Secular philosophy has engaged in a heated debate about who/what might legitimately be included in theories of social justice. In order to arrive at a reasoned position, it is

13. Deane-Drummond, *Ecotheology*, 116–18.

14. Deane-Drummond, "The Bible and Environmental Ethics."

15. Deane-Drummond, *The Ethics of Nature*.

important to distinguish between agents who are *distributors* of justice and *recipients* of justice. This allows for a reasoned inclusion of non-human species (or ecosystems) as those that are also recipients of distributive forms of justice.

BOX 4: VARIETIES OF JUSTICE

General justice, also sometimes called *contributive justice* between the individual and state that is orientated to the common good can be compared with *distributive justice* between the state and the individual and *commutative justice* between individuals. If we broaden the scope of justice to include the non-human community as recipients of justice in the way I have suggested above, then this would not only impact on distributive justice, but also commutative justice would impinge on our own individual lifestyles and choices. General justice is also particularly relevant when applied to multinational companies that are presently making an even greater contribution to the economy than many nations. Compensatory justice refers to compensation after environmental damage, such as the kind of damage wrought through the Dakota pipeline construction discussed in BOX 1, though normally this just applies to the impact on human livelihoods understood more narrowly rather than religious sensibilities or the ecosystem.

For example, where multinational companies are making an inordinate profit through biotechnology patents, then general justice would insist on making payments by taxes. Of course, this also raises the issue of whether other interests are being denied through such activities, in addition to purely economic gain. Where the growth in biotechnology has had a negative environmental impact, for example, through loss of biodiversity, or through exploitation of "wild" reserves in poverty stricken regions, then *compensatory* justice should also apply. Multinational companies taking advantage of laxer environmental laws in one jurisdiction in order to gain trade advantages need to be held to account.

THEOLOGY AND THE ECONOMY

How far does the market economy reflect Christian ideals? Harvey Cox believes that the market economy undercuts the traditional meaning of the land in human history:

> For millennia of human history, land has held multiple meanings for human beings—as soil, resting place of the ancestors, holy mountain or enchanted forest, tribal homeland, aesthetic inspiration, sacred turf. The Market transforms all these complex meanings into one: land becomes real estate; there is no land that is not theoretically for sale, at the right price.[16]

From this perspective, those who assimilate and endorse fully a market economy as expressive of Christian religion, the so-called prosperity gospel, are deeply mistaken. Pope Francis's encyclical *Laudato si'* was equally sharply critical of the market economy and developed Pope Benedict XVI's transformation of economics to that of personal gratuitousness rather than rest on the unseen "magic" hand of the market.[17] This is also a significant shift away from an emphasis just on human rights, so "relationships of gratuitousness, mercy and communion" are more "fundamental" than "relationships of rights and duties."[18] His primary intention is to reform in the first place how human beings relate to each other, rather than

16. Cox, "Mammon and the Culture of the Market," 277.
17. Deane-Drummond, "Technology, Ecology and the Divine."
18. Benedict XVI, *Caritas in Veritate* §7.

fixing the problem after a breakdown through a demand for rights. Although he does not spell this out, it would mean that in situations where multinational companies are, for example, bent on extracting ore, or developing genetically modified crops, due account of the needs of the local population and the natural environment would be taken into account.

Economic thinking based purely on the market tends to assume that there are unlimited natural resources, which clearly there are not. The idea that an environmental good can be given a monetary value equivalent, in the manner of some more environmentally inclined economists, is one resisted by many philosophers, mostly because ecological concern reflects human values as citizens, whereas the economy presupposes preferences as consumers, hence giving monetary value to environmental goods is a category mistake.[19] People habitually refuse to give specific monetary values for environmental protection once it is couched in the language of cost-benefit analysis, such as questions about willingness to pay (WTP) for particular environmental goods that do not have a specific economic value in the market place.

Classical prudence or practical wisdom means taking counsel, judging, and acting according to the common good. It also includes insight, memory of the past, awareness of present circumstances, foresight, and reasoning. It is relevant for environmental decision-making as it resists a narrow allocation of monetary value to environmental goods through cost benefit analysis (CBA).[20] Political prudence implies a process of *deliberation*, and in this

19. Smith, *Deliberative Democracy*, 29–49.

20. Deane-Drummond, "Wisdom, Justice."

sense comes closer to the political models of deliberative democracy suggested by others.

BOX 5: ENVIRONMENTAL INDULGENCES

Putting cash values on environmental goods through other means such as compensatory eco-taxes also encourages the idea that it is all right to pollute as long as compensation is paid, and it is all right to develop strategies that are unsustainable. Some economists have argued that paying for what is known as "externalities," that is unwanted environmental damage, is a more favorable alternative compared with legislation that then demands penalties for violation. Yet such compensation smacks of what Robert Goodin has called environmental "indulgences"; for it is selling what is not yours to sell, it is selling what cannot be sold, and it is rendering wrongs right. It would therefore be preferable to have stronger positive law against violation, which at least names the action as wrong, rather than something that can be compensated for through incentive payments to others that are harmed.

Goodin, "Selling Environmental Indulgences."

Are there strategies for achieving alternative economies that are more sustainable compared with the current market economy? John Cobb in a classic text has argued for a communitarian approach through the establishment of small regions that are self-sufficient in terms of basic survival needs through the decentralization of the economy.[21] Such regions would impose tariffs on goods that are produced elsewhere where the wages are low, or where there is pollution or poor working conditions generally. He believes that regions may be national in some cases, but in others represent villages that culturally can operate in a self-sufficient way.

21. Cobb, "Towards a Just and Sustainable Economic Order."

ENVIRONMENTAL JUSTICE MOVEMENTS

There is a need to pursue alternative strategies simultaneously. In the first place, the situation of environmental injustice understood as unfair distribution of environmental harms and goods needs to be addressed urgently, prior to any desirable revolution in the economy. A purely communitarian approach might also lead to an insularity toward other, less fortunate communities. Extreme poverty, social exclusion, and environmental injustice appear in tandem in communities all over the world. This is one reason for the environmental justice movement, but those interested in environmental ethics should not ignore such action on behalf of these communities.

Could there be *intermediate mediating strategies* between the market-led system that has dominated the political economy to date and more radical idealistic alternatives? Such mediating strategies need to be viewed as interim emergency measures prior to more widespread change, rather than a reformist strategy per se that is content merely to stay with such reform measures. Bringing environmental questions to bear via environmental justice helps to put ecological issues onto the political agenda.

Interim reformist measures are only likely to be satisfying in the short term for emergency purposes, for they leave the basic model of the economy intact. The movement toward concern for future generations and nonhuman species is more likely to be politically achieved in a democratic society through constitutional changes. One of the difficulties of a global market economy is that it serves to undermine the economic authority and political power of individual nation states.

DEVELOPING THEOLOGICALLY INFORMED ECOLOGICAL ETHICS

I will suggest here just a few approaches that might be helpful to develop in order to elaborate a new way of conceiving ecological ethics that has a Christian "soul" or ethos:

(1) *Liturgical transformation*. Those ancient liturgies that insisted on creaturely participation in the praise and joy in God need to be reclaimed and celebrated, along with new ones designed for a contemporary context. Alongside this, we need to find appropriate means to confess our guilt to one another and to God, not as a way of removing responsibility but in order to acknowledge our share in the failure of human societies to live up to environmental ideals.

(2) *Global and local ecclesial responsibility.* Once environmental consciousness seeps into our religious experience on a regular basis, we can expect changes in attitudes that are desirable on a local and global scale. Those churches that have a universal reach have a particular responsibility to represent a Christian theological view in the *global* public sphere. But the ecumenical mandate of all Christian communities is to work together at a *local* level to build ecologically responsible forms of flourishing.

(3) *Practical steps in individual responsibility.* A first simple step is through being more self-conscious about the kind of food that we eat and its ecological footprint. There are, of course, other decisions where we can make a difference, including being more aware of the overall energy consumption. Some may

feel a sense of responsibility to work at ameliorating structural sin by working with non-government organizations or other forms of political advocacy.[22]

(4) *Building a collective conscience.* By this I mean an awareness of what the communities in which we are placed assume as the norm for moral action. Once we are aware of these, there will be some norms that need to be challenged. Collective conscience is more than just collective consciousness, as it is about moral norms that are shared at different community levels. Building a collective conscience that is self-consciously more environmentally aware is essential if complex problems such as climate change are going to be addressed. Christians have something important to contribute to what conscience means and how to foster sensitivity to it at individual and collective levels.[23]

But it is the environmentally sensitive *practices* of Christians and their witness to a different kind of lifestyle that will, perhaps, speak louder than words in building a Christian ethos in a way that is both faithful to Christian creeds, yet at the same time, communicates to those with other religious faiths or none the central importance of ecological responsibility.

22. For some examples of this practice see Deane-Drummond and Bedford-Strohm, eds., *Religion and Ecology.*

23. I have discussed this idea further in Deane-Drummond, "A Case for Collective Conscience."

REFLECTION

1. What are the ways in which you think that Pope Benedict XVI's economy of gratuitousness could be put into practice in your local area?
2. Do you think that the category of ecological sin provides motivation for action or not?
3. Are there examples of environmental injustice in your community? What action could be taken?
4. Can you think of ways of implementing approaches to developing theologically informed ecological ethics named 1–4 above and add to this list?

POSTSCRIPT

Gaining Ecological Wisdom

One of the most agonizing aspects of dealing with questions about ecology and sustainability is that even at a secular level these questions are extremely difficult, if not impossible, to solve. Indeed, the idea that there might be a technical fix or "solution" is misguided if it presumes an overconfidence in technology or that only an elite has the capability to make decisions that impact on the planet as a whole. Eco-modernists, for example, have undivided faith in human abilities to solve complex problems such as climate change through technological innovation. Some issues, such as examples of environmental injustices toward first nation or other oppressed peoples are relatively straightforward. It seems obvious what should be done and for what reason. Climate change is much more complicated ethically, since there are complex global issues to deal with, as well as a failure to recognize the harm caused by actions such as mobility or heating a home that are in themselves and on their own goods to be sought. In these scenarios Pope Francis is, it seems to me, correct to focus on the way global culture has idolized technology. What used to seem like a luxury has now become a right. The technocratic paradigm is not simply about technologies as such, but the way they have crept into our psyche almost without recognizing the changes in lifestyle that fall in their wake. No one wants to go back to a time when there were few, if any, of the benefits of modern medicine or other communication tools that make modern living so much easier, such as that provided by internet technologies.

The difficult discernment is working out how much consumption is really necessary for a flourishing human and planetary life, and for what reasons. The global economy is still predominantly tied to a market model where ecological and even social harms are considered mere "externalities" in working out the specific costs of manufacturing a given product or providing a specific service. The radical shift in economic theory proposed by Pope Benedict XVI and elaborated by Pope Francis is one where there is recognition that the economy itself is a human enterprise, and therefore capable of being changed by actors committed to taking a different course. Ecological economics is not impossible to achieve, but it is still swimming against the tide of global finance and standard market measures. Further, will global economics make the transition to a carbon free economy in a way that also sheds its attachment to technological development as the proposed solution to complex human problems? In other words, the temptation for the new economy is to seek technologies that address the problems of climate change, rather than seeking changes in the human mind and spirit.

I believe that the ancient wisdom that arises in ecotheology can make a genuine contribution to the societal and individual changes that are necessary to deal with ecological questions responsibly. The Catholic tradition has called this ecological conversion. Theology, at its best, considers the whole and, given that its ultimate authority is God, *theo*centrism, is not, theoretically at least, attached to any specific humanistic framework. Anthropologists, of course, will protest that theology arises out of a specific and narrowly defined cultural tradition. In the light of this insight, theologians need to recognize that there are other players that need to be consulted when coming to difficult decisions, even if they want to remain true to their own faith based standpoint. The specific wisdom of Christian traditions therefore needs to be joined up with the wisdom arising from other religious faiths, and those who are non-believers but share the concerns in bringing about a better world, both for our own generation and future generations.

Practical wisdom or prudence is critical at all levels, not just in making local decisions by individuals, families, or local community groups, but also on a larger scale of nations and international communities. Navigating the demands of the individual in relation to community, of nation in relation to other nations, is one of the challenges to be faced when dealing with the global commons. Subsidiarity as a principle is helpful, that is, making a decision where possible at the lowest level of organization. But there are important issues to be considered at the global level that attention to local contexts may skirt around. A useful map of the kind of issues that need to be addressed at the very broadest level arises in the *planetary boundaries* model, which addresses nine different planetary risk thresholds simultaneously, including, of course, climate change, but also nitrogen cycles, biodiversity, and ozone layers, among others. Scientific consensus on the existing or threatened violations of planetary boundaries necessary for the stability of the earth as a whole system mean that emboldened attempts to halt habitat disappearance, including biodiversity loss, seem doomed, at least in the long run.[1] Kate Raworth from the development agency Oxfam has pressed for an inclusion of vulnerable people in this schema, so that there is a safe place for humanity to live existing within these boundary limits.[2] This integration of the needs of people with the planet has been a characteristic of Catholic social teaching from its inception, yet what such models do not address are difficult questions of conflicting interests.

Planetary boundaries and geological concepts that have escaped from their original usage in geological science such as the Anthropocene have a leveling out effect on human responsibility. In one sense it is true that all humans collectively are responsible for the damage that we have done. But at another level this is profoundly false, since there is a disproportionate impact of those that are consuming the most. The Anthropocene is also undermining attention to nature as the other to which we respond, a concern that is at the heart of most secular environmental ethics, and which inspired wilderness American activists such as John Muir and Henry Thoreau. A Christian view treats the otherness of nature as given by God, so the earth is a Gift to humanity. The Anthropocene crowds out this otherness and giftedness and differential responsibilities

1. Rockström et al., "A Safe Operating Space for Humanity," 472–75; Rockström et al. "Planetary Boundaries," 32. See also Steffen et al., "Planetary Boundaries."

2. Raworth, "A Safe and Just Space for Humanity," 1–26.

between different groups. In my view Pope Francis is correct to name the priority of the basic need for water and other goods necessary for human survival as having ethical priority over biodiversity loss. But in many cases those survival needs can be met without threatening the welfare of other creatures.

The treatment of animals in concentrated feeding lots that are widespread in the USA is just one example where priorities have gone awry and there is a profound lack of wisdom. Choosing not to eat food sourced from animals reared in such conditions is just one of the small actions of protest that individuals and communities can make. It is a disappointment in my view that connections such as this are not yet recognized in the institutions of which we are part that claim to care for God's creation. Indeed, one of the difficulties for any theological approach is moving from head to heart, from the inner world of academe to that of practice and perhaps even protest. Is civil disobedience the only real way forward in order to bring about wide scale social change, as environmental activist and social scientist Bill McKibben passionately believes?[3] The Catholic tradition has generally avoided making specific recommendations for radical peaceful protest. But as the coming decades dawn, the situation may reach such a level of seriousness that such action is necessary as a last resort. Political establishments will do well to pay heed to the risk of instability and even the threat of violence as the problems become more acute.

Finding the roots of our compassion for others and for the earth is, as this book has argued, premised on belief in a God who loves the world and who everyone can experience in the natural world around us. God's presence is

3. McKibben, *Oil and Honey*.

everywhere, as the ancient writers from Basil of Caesarea through to more contemporary scholars such as Denis Edwards have affirmed. That divine presence in the world makes that world sacred, rather than divine in the Christian tradition, so that the Spirit of God is also the Spirit of Creation. Most scientists and biologists are richly attuned to the wonder in the natural order, but from a Christian perspective, many will miss the source and reason for why that awe rises up in the human spirit. Christian theology provides insight as to where that awe originates, namely, in the love of God for creation. Further, the special task given to humanity is one of taking responsibility, rather than avoiding the demands laid upon us as human persons made in the image of God. The human cruelty that is evident all around us arises from broken relationships with God, each other, and the natural order.

An ecological pneumatology will seek to foster the movement of the Spirit in a different direction, one that works toward healing of broken relationships with each other and the earth. Where the Spirit may be blowing is something that cannot be contained or controlled. But each person is under a serious obligation to discern what that task might be. We cannot all be working to try and remedy precisely the same facet in this juggle of knots that makes up climate change, environmental degradation, and human and planetary disease. Becoming conscious of what our gifts might be in this field and putting that into practice is the task of a lifetime. It is my hope that this book will sow a few seeds that may take time to mature and grow, but however small, like mustard seeds, if planted by God then they will be impossible to suppress. Ecotheology needs to be woven into the rhythm of life: our daily practice, our liturgical celebrations, and our family meals. The next generation will look to see how far we have been faithful to that call and they, like the creatures that will no longer see another dawn, will wait expectantly for our reply.

APPENDIX

Christian Environmental Activism

Many of my students find that really engaging in depth with ideas relevant for ecotheology is transformative, leading to a new energy to want to make a difference in the world as it hovers on the threshold of dramatic and irreversible change. This generation has a huge responsibility that can be overwhelming. Yet there are practical things that can be done, agencies that are working specifically in building a better world for both people and planet. Ecotheology, unlike pure environmentalism, is concerned with people and creatures as part of the wondrous gift of God the Creator.

The following web sites are those that give more information about Christian environmental activism, or faith based activism that is ecumenical in scope.

CHRISTIAN ENVIRONMENTAL ACTIVIST ORGANIZATIONS

A Rocha, http://www.arocha.org/en/

Creation Justice Ministries, http://www.creationjustice.org/

Evangelical Environmental Network, http://www.creationcare.org/

Interfaith Power & Light, http://www.interfaith-powerandlight.org

Lutherans Restoring Creation, http://www.lutheransrestoringcreation.org/

Restoring Eden: Christians for Environmental Stewardship, http://restoringeden.org/

FAITH-BASED ENVIRONMENTAL ACTIVIST ORGANIZATIONS

Alliance of Religions and Conservation, http://www.arcworld.org/

Earth Ministry, http://earthministry.org/

GreenFaith, http://www.greenfaith.org/

Interfaith Center for Sustainable Development, http://www.interfaithsustain.com/

National Religious Partnership for the Environment, http://www.nrpe.org/

POPULAR AND PUBLIC NETWORKS

Catholic Climate Covenant, http://www.catholicclimatecovenant.org/

European Christian Environmental Network, http://www.ecen.org/

ORGANIZATIONS THAT INCLUDE ENVIRONMENTAL PROGRAMS

Catholic Agency for Overseas Development, One Climate, One World Campaign, http://cafod.org.uk/Campaign/One-Climate-One-World/About-the-campaign

Catholic Relief Services, http://www.crs.org/

Catholic Rural Life, https://catholicrurallife.org/

Green Jesuit, http://greenjesuit.org

Jesuit Volunteer Corp, http://www.jesuitvolunteers.org/

United States Conference of Catholic Bishops—Environmental Justice Program, http://www.usccb.org/issues-and-action/human-life-and-dignity/environment/index.cfm

ACADEMIC COLLABORATIONS

Yale University's Forum on Religion and Ecology, http://fore.yale.edu/

Society of Conservation Biology—Religion and Conservation Biology Working Group, https://conbio.org/groups/working-groups/religion-and-conservation-biology

Society of Conservation Biology—Best Practices for Religious and Indigenous Community Interaction, https://conbio.org/publications/scb-news-blog/best-practices-for-religious-and-indigenous-community-interaction

Torreciudad Declaration between climate scientists, religious leaders and theologians: http://www.declarationtorreciudad.org/

GLOSSARY OF KEY WORDS IN ECOTHEOLOGY

agrarianism, as the name implies, signifies a particular approach to farming and agricultural practice, often in tune with local traditions tested over generations, rather than one that uses, for example, genetic technology, but whose practice then informs a particular social philosophy of life as such.

androcentrism is a focus on the worth of humans, specifically men, to the exclusion of women and other creatures. It is a particular variant of anthropocentrism.

animal rights is a social and political movement that claims animals have individual moral worth by analogy with human rights and therefore each animal has its own special dignity and are subjects of a life.

animal studies is a broad field taking in critical social sciences and political discussions about the place of animals in human societies. It is distinct from animal behavior since the latter is a natural scientific study, while animal studies aligns more with social sciences and in some cases religious studies approaches.

Anthropocene is a new geological era under discussion by geologists conscious of the unprecedented impact of human beings on planet earth and planetary systems. While the start date of the Anthropocene is controversial, many associate it with the rise of industrialization or the invention of the steam engine.

anthropocentrism is a philosophical approach which concentrates in an exclusive way on human action, structures and cultures. It is closely related to the idea of human exceptionalism, the belief that humans are more important than any other creature.

biocentrism is the belief that all biological beings are morally considerable and have value and is normally intended to mean the life system on planet earth as a whole, rather than its separate components. Sometimes those who hold this view are accused of eco-fascism, as the value of individual creatures, including humans, are downplayed.

biodiversity describes the variety of different life forms on the planet normally under the banner of speciation. Biodiversity is currently under threat due to the loss of habitat or direct killing by poaching, illegal trade and legitimate practices that indirectly impact on the well being of different species.

biopolitics refers to the intersection between political theory and the biological realm, though in postmodern thought it has also been used to refer to the dominance of political power over life forms.

bioregionalism refers to localized practices according to ecological categories, so, for example, what brings different communities together according to their ecological setting, such as a watershed for example.

biosemiotics studies signs or codes in the biological realm and joins the field of semiotics with that of biology.

biosphere is the sum total of biological species on planet earth.

capabilities approach refers to a theory of justice according to specific goals of flourishing, either in alignment with basic human rights approaches, or in accordance with the desires of that community through collective decision making. It can be adapted to apply to animals as well in accordance with human perception of their particular needs.

carrying capacity of the earth usually refers to the ability of the planet as a whole to support life indefinitely, though it can be used a more localized way with specific species in specific regions. In the context of population growth it refers to the maximum number of people that could live on the planet given the available resources without degrading that planet for future generations.

charity in the classic tradition refers to love of God and neighbor.

Christic is a reference either to the cross of Christ or Jesus Christ and is a descriptive term that is intended to highlight links between Christology and other areas under consideration.

Christology is the study of Jesus Christ in terms of his significance for salvation, soteriology, future hope, eschatology, and theological anthropology. In its broadest sense it also refers to his significance for creation as such through a Trinitarian approach.

climate change deniers are those who resist, often for political reasons, data which provides evidence for climate change, such as melting ice caps, warmer global temperatures and unpredictable weather patterns.

commutative justice refers to what is owed following agreements between individuals according to basic principles of fairness. Private organizations are often treated as individuals in a legal system.

compensatory justice refers to what a collective body or other party owes to an individual when they have been injured or their legitimate claims have been violated.

consequentialism is an ethical theory that deals with consequences of particular actions as a way of judging whether an action is right or wrong.

conservation is a field concerned with the protection of wild life in its different forms, along with ecosystems, landscapes and habitats.

constitutive justice refers to an assessment of the justice of a particular communal system of justice; so, for example, slaves were considered outside the requirements of justice until that system of justice was questioned through constitutive justice.

contributive justice is the responsibility that each member of community has to a collective, usually a nation or state.

cosmic communion refers to the intimate relationship between different entities in the cosmos, sometimes thought in theological terms to mirror the Trinity.

cosmos is the broadest sweep of the known universe or universes, so the sum total of the material world.

creatures often refers to living organisms that are understood to be created by God, but can also in ancient texts refer to the material world as such.

cruciform is a descriptive adjective pointing to the relationship between what is observed in the world and the cross of Christ hence his passion and death.

deep ecology is a political movement that lays out a platform for political activism and gives priority to ecological wellbeing, sometimes over and above human wellbeing.

deep incarnation is a way of expressing the idea that when God became incarnate in the person of Jesus Christ, this had significance for the whole cosmos and not just human beings.

deontological ethics refers to basic principles by which ethical decisions are made, such as the principle of duty to act justly toward one's neighbor.

distributive justice is the duty of the state to act fairly toward the individual.

dominion is the relationship between humanity and the natural world normally considered in terms of stewardship or caregiving.

earth Bible is a particular reading of the Bible that pays attention to ecological issues, either in the text itself or through paying particular attention to eco-justice principles.

eco-criticism is a critical reading of literary texts that place particular significance on environmental questions.

eco-justice principles are those drawn up in the earth Bible team in order to read the Bible in a particular way, so, for example, the principle of voice, that the earth has agency, the principle of intrinsic value, the principle of mutual custodianship, the principle of interconnectedness and the principle of resistance, that the earth itself is subject to injustices from human oppressive actions and will actively resist those actions.

ecofascism is directed against radical ecology supporters that seem to support the good of the earth at the expense of individual human lives. The association with fascism is intended to undermine the case for earth care.

ecofeminism is a more recent movement within feminism sometimes called the "third wave," and it takes a number of different forms, depending on the specific political and religious alliance of the feminists involved.

ecological crisis is the belief the earth and its living components are being damaged irreversibly through human acting within a complex system whose problems are very difficult to solve.

ecological economics is a re-crafting of economics so that it takes better account of ecological harms, rather than just treating these as extraneous "externalities" that are an unfortunate side effect of the market economy.

ecological justice is the broadest concept of justice for the earth that includes both humans and other creatures and what is their due.

ecological restoration is the belief that some environments after they have been damaged can and should be restored to a former state that more closely aligns with an ideal historical one.

ecology is the study of living organisms and their interrelationships.

ecosystem is the way in which different organisms interact in a particular way forming a specific system.

encephalisation refers to the increase in relative size of the brain over the course of evolution.

environmental humanities is a sub-field of different humanities subjects which are all concerned with and pay attention to environmental issues and problems.

environmental justice concerns what environmental goods are owed to particular people, and those who suffer environmental injustice are also liable to suffer other forms of injustice as well, such as racism and extreme poverty.

eschatology refers to the future history of the universe, planet and humanity according to Christian hope and expectation, but that future is also considered to break into the present in what is known as "realized" eschatology.

essentialism refers to some fixed set of characteristics that are needed for identity. It is often used in a derogatory manner, for example, an essentialist view of women in the context of the way their identity is restricted to particular characteristics drawn up by men.

ethno-elephantology is the study of elephant behavior using the tools of ethnography used in human communities.

ethnology is the detailed study of human behavior by anthropologists who live alongside that community.

ethology is the detailed study of animal behavior.

evolution is the process of change in the earth and living organisms over long periods of history.

extinction normally refers to the complete loss of a particular biological species or culture.

future primitive is the view that the ideal future also includes a backward look at societies prior to industrialization and techniques of modern agriculture.

Gaia refers to the earth Goddess of ancient Greece. It has also been adopted by maverick scientist James Lovelock to refer to a theory which describes the sum total of the earth system and its relatively stable atmosphere (at least until recently) that is stable as a result of the regulating effect of the biota.

general or ***legal justice*** refers to the broadest category of justice and provides reasons for what is due to whom.

genome is the genetic information summed up in any particular organism.

globalization is the political and social processes through which goods and services are exchanged at a global level.

GMOs are genetically modified organisms which refers to the way in which genetic technology is used to change or modify the phenotypic characteristics of a particular organism in a way that may or may not be heritable.

God's body is the belief that the earth as such is the body of God, even though God is greater than this body.

green is the political inclination toward environmental activism.

green anarchists are those committed to extreme action even if it is against the law in for environmental ends.

habitat is the network of associations by each living organism that contribute to its overall function. If a habitat is lost that species or community is likely to disappear unless they are resilient to change.

hermeneutics is a theory of interpretation, and for Christians this is normally a particularly significant text such as the Bible.

human ecology has a specific meaning in Catholic social thought that goes beyond its meaning in the social sciences. For social scientists it refers to the living context in which humans dwell, but in Catholic social teaching it refers to the primacy of the human within that context in so far as it is given by God the Creator.

imago Dei refers to the divine image as spoken about in the book of Genesis. It is what makes humans distinctive, and may say unique compared with all other planetary creatures. Theologians who believe that image bearing has fostered unwarranted anthropocentrism may either dispense with this term or widen its use to be more inclusive of other creatures.

imperialism is a political national order where there the dominant ruling class acquires territory belonging to other nations.

indigenous refers to ancient local populations that were present before imperialist acquisition when referring to humans, and when referring to biological species it is those that were present prior to any invasion of other species. Of course, the definition of indigenous depends on how far in history one goes back, but normally indigenous communities are recognizable from their much longer history of association with the land over and above any imperialist or colonizing ones.

inherent value is value given by humans to particular creatures or places.

instrumental value is value only so far as it is useful for humans.

integral ecology is an integrated approach to environmental problem solving that includes economic, social and biological dimensions. It has been used in Catholic social thought to refer to the broader implications of natural law. In the work of Pope Francis it also refers to integral development and a broader vision of reality that is given by God.

intrinsic value is the good of each creature in and of itself.

landscape is the land in its visual and aesthetic aspect as shaped by particular activities of different organisms and cultural interventions.

liminal refers to an in between space, and can also imply the boundary between material and spiritual realities.

market economy is the political system of exchange of goods according to its monetary value. The global market refers to the price of goods when leveled through international trade agreements.

metanoia is the change of heart and mind, a turning around.

militant ecology activism is the political inclination to act in a radical sometimes illegal way for the sake of preservation of a particular ecology that is valued as a highest good.

multispecies are the association between different species in a communal set of relationships.

myth is a world view or story about the world that may have specific religious undertones.

natural law in its earliest definitions refers to seeking the good and avoiding evil, where goods are considered through the baseline of the flourishing of all life, and where the natural world and humans in particular are also thought to be imbibed with reason.

nature is the biological realm consisting of all humans and other creatures, though it can also refer to that which is other than humans, or even human nature as such.

nature writing are those literary scholars such as John Muir who developed classic texts on writing about their experiences in and with the natural environment.

neo-cortex is the upper part of the brain thought to be associated with intelligence and human reasoning.

new creation story is a grand narrative that reaches back to the beginning of the universe and seeks to describe how God created the world while weaving in particular scientific imagery and insights.

ontological refers to the metaphysical category of what a living thing is in and of itself as well as the kinds of things that have existence.

place is the physical location in which humans and other animals dwell. It becomes home when humans craft it in particular ways to suit their needs.

planetary boundaries are those physical limits thought to be characteristic for the right functioning of the earth as a system. There are nine planetary boundaries that could potentially be breached.

Pneumatology is the study of the action of the Holy Spirit in the creation and more traditionally in the lives of Christians and other creatures susceptible to prompting.

practical wisdom is a way of making decisions according to a broadly Aristotelian framework that includes deliberation, judgment and action.

preservation is concerned with keeping something alive.

Rawls' difference principle is a theory of social justice that claims that an action is good if it benefits the least fortunate at the same time as delivering goods for the individual involved.

religious environmentalism is environmental action inspired by religious motivations.

scale normally refers to different biological categories that are the focus of study, ranging from the micro-scale, through to the level of the organism, community, niche and canopy.

scientism is the belief that scientific reasoning offers the only way to perceive the world so that scientific ways of knowing become formative for all other decision making.

species are those organisms which are capable of reproduction of specific kinds like their own. Species was popular as a term in early biological studies, but many biologists are now more skeptical because of the fluid genetic exchange that is possible between species.

stewardship refers to that theory of environmental responsibility that gives precedence to human management of the planet and its resources in ways that are intended for the benefit of the system as a whole as well as individual goods. It is a vague term that can mean different things depending on context, but it is often popular in ecclesial literature because of its stress on human action.

sublime communion is the ultimate goal toward which the planet is tending according to Christian eschatology.

sustainability refers to the persistence of environmental, social and material goods in future generations. The term has been co-opted by different groups so that its practical meaning varies in terms of the extent of environmental priority. At its best it signals the difficulty and challenge of working across different fields of study in trying to solve difficult environmental problems at the political, social, ethical, environmental and religious levels.

symbiotic refers to the pattern of interaction between organisms where both mutually benefit from the association.

temperance is a habit of mind or disposition that is determined not to take more than needed for healthy existence, and so is closely connected with self restraint and lack of greed.

theo-centrism is the perspective that God is the center of reality rather than humans or other creatures.

theodicy is the philosophical challenge of trying to explain how a claim can be made that God is good when there is evil and suffering in the world.

theological virtues refer to the virtues of faith, hope and love.

trinitarian refers to the pattern of existence that mirrors that found in the relationship between the persons in the Godhead, namely, Father, Son and Spirit.

utilitarian ethics refers to a theory of ethics that is concerned with what leads to greatest happiness or utility according to the greatest good for the greatest number.

utopia is a belief in an idealized future.

virtue ethics is a system of ethics that it based on the importance of considering what is right in terms of the way humans act as agents, rather than simply consequences or principles.

wild justice is the application of language of justice to the way social animals associate with each other, particularly according to rules of play.

wilderness preservation is focused on those parts of the environment that seem to have less interference by human beings, and so be more "pristine."

wisdom is an intellectual virtue that refers to the ability to form right relationships with everything, including a relationship with God for those who are theists.

wonder is an emotional reaction to contact with either beauty or the repulsive, and in ecotheology that means the natural world as such, or it may be associated with a search for knowledge where the boundary of that knowledge is unknown.

BIBLIOGRAPHY

Alaimo, Stacy. *Undomesticated Ground: Recasting Nature as Feminist Space*. Ithaca, NY: Cornell University Press, 2000.

Ammons, Elizabeth, and Modhumita Roy. *Sharing the Earth: An International Environmental Justice Reader.* Athens: University of Georgia Press, 2015.

Anderson, Paul N. *The Riddles of the Fourth Gospel: An Introduction to John.* Minneapolis: Fortress, 2011.

Balthasar, Hans Urs von. *Theo-drama: Theological Dramatic Theory*, Volume IV: *The Action.* Translated by Graham Harrison. San Francisco: Ignatius, 1994.

Barker, Margaret. *Creation: A Biblical Vision for the Environment.* London: T. & T. Clark, 2010.

Bauckham, Richard. *Bible and Ecology: Rediscovering the Community of Creation.* Sarum Theological Lectures. Waco, TX: Baylor University Press, 2010.

Baynes-Rock, Marcus. *Among the Bone Eaters: Encounters with Hyenas in Harar*. University Park: Pennsylvania State University Press, 2015.

———. "Life and Death in the Multispecies Commons." *Social Science Information* 52 (2013) 210–27.

Benedict XVI, Pope. *Caritas in Veritate.* London: Catholic Truth Society, 2009.

Bergmann, Sigurd. *Creation Set Free: The Spirit as Liberator of Nature*. Grand Rapids: Eerdmans, 2005.

Boff, Leonardo. *Cry of the Earth, Cry of the Poor*. Translated by Philip Berryman. Ecology and Justice Series. Maryknoll, NY: Orbis, 1997.

———. *Ecology and Liberation: A New Paradigm*. Translated by John Cumming. Ecology and Justice Series. Maryknoll, NY: Orbis, 1995.

Brague, Rémi. *The Wisdom of the World: the Human Experience of the Universe in Western Thought*. Translated by Teresa Fagan. Chicago: University of Chicago Press, 2003.

Brown, William. *Wisdom's Wonder: Character, Creation and Crisis in the Bible's Wisdom Literature*. Grand Rapids: Eerdmans, 2014.

Cahill, Lisa. "The Environment, the Common Good and Women's Participation." In *Theology and Ecology across the Disciplines: On Care for Our Common Home*, edited by Celia Deane-Drummond and Rebecca Artinian Kaiser. London: Bloomsbury, 2018, *in press*.

Carson, Rachel. *Silent Spring*. Boston: Houghton Mifflin, 1962.

Cleland, Elsa E. "Biodiversity and Ecosystem Stability." *Nature Education Knowledge* 3 (2011) 14.

Cobb, John B. "Towards a Just and Sustainable Economic Order." In *Environmental Ethics: An Anthology*, edited by Andrew Light and Holmes Rolston, 359–70. Blackwell Philosophy Anthologies 19. Oxford: Blackwell, 2003.

Conradie, Ernst. "Confessing Guilt in the Context of Climate Change." In *Ecological Awareness: Exploring Religion, Ethics and Aesthetics*, edited by Sigurd Bergmann and Heather Eaton, 77–96. Berlin: LIT, 2011.

Conradie, Ernst, et al. *Christian Faith and the Earth: Current Paths and Emerging Horizons in Ecotheology*. New York: Bloomsbury T. & T. Clark, 2014.

Cox, Harvey. "Mammon and the Culture of the Market: A Socio-Theological Critique." In *Liberating Faith: Religious Voices for Justice, Peace, and Ecological Wisdom*, edited by Roger S. Gottlieb, 274–83. Lanham, MD: Rowman & Littlefield, 2003.

Dalton, Anne Marie, and Henry Simmons. *Ecotheology and the Practice of Hope*. SUNY Series on Religion and the Environment. Albany: State University of New York Press, 2011.

Daneel, Marthinus L. "Earthkeeping Churches at the African Grass Roots." In *Christianity and Ecology: Seeking the Well Being of Earth and Humans*, edited by Dieter T. Hessel and Rosemary

Radford Ruether, 531–52. Cambridge: Harvard University Press, 2000.

Davis, Ellen F. *Scripture, Culture and Agriculture: An Agrarian Reading of the Bible.* Cambridge: Cambridge University Press, 2009.

Deane-Drummond, Celia. "Beyond Humanity's End: An Exploration of a Dramatic versus Narrative Rhetoric and Its Ethical Implications." In *Future Ethics: Climate Change and Apocalyptic Imagination*, edited by Stefan Skrimshire, 242–59. London: Continuum, 2011.

———. "The Bible and Environmental Ethics." In *The Oxford Handbook of Bible and Ecology*, edited by Hilary Marlow and Mark Harris. Oxford: Oxford University Press, 2018, *in press.*

———. "The Birth of Morality and the Fall of Adam Through an Evolutionary Inter-species Lens." *Theology Today* 72 (2015) 182–93.

———. "A Case for Collective Conscience: Climategate, COP-15 and Climate Justice." *Studies in Christian Ethics* 24 (2011) 5–22.

———. "Catholic Social Teaching and Ecology: Its Promise and Limits." In *Fragile World: Ecology and the Church*, edited by William Cavanaugh. Eugene, OR: Cascade Books, 2018, in press.

———. *Christ and Evolution: Wonder and Wisdom.* Theology and the Sciences. Minneapolis: Fortress, 2009.

———. "Creation." In *Cambridge Companion to Feminist Theology*, edited by Susan Frank Parsons, 190–207. Cambridge Companions to Religion. Cambridge: Cambridge University Press, 2002.

———. *Creation through Wisdom.* Edinburgh: T. & T. Clark, 2000.

———. "Deep Incarnation and Eco-justice as Theodrama." In *Ecological Awareness: Exploring Religion, Ethics and Aesthetics*, edited by Sigurd Bergmann and Heather Eaton, 193–206. Studies in Religion and the Environment 3. Berlin: LIT, 2011.

———. *Ecotheology.* London: Darton, Longman & Todd, 2008.

———. *The Ethics of Nature.* New Dimensions to Religious Ethics. Oxford: Wiley/Blackwell, 2004.

———. "Joining the Dance: Catholic Social Teaching and Ecology." *New Blackfriars* 93 (2012) 193–212.

———. "*Laudato Si'* and the Natural Sciences: An Assessment of Possibilities and Limits." *Theological Studies* 77 (2016) 392–415.

———. *Re-Imaging the Divine Image: Humans and Other Animals.* Kitchener, ON: Pandora, 2014.

———. "Technology, Ecology and the Divine: A Critical Look at the Rising Tide of New Technologies through a Theology of Gratuitousness." In *Just Sustainability: Technology, Ecology, and Resource Extraction*, edited by Christiana Z. Peppard and Andrea Vicini, 145–56. Catholic Theological Ethics in the World Church 3. Maryknoll, NY: Orbis, 2015.

———. "Who on Earth Is Jesus Christ?: Plumbing the Depths of Deep Incarnation" In *Christian Faith and the Earth: Current Paths and Emerging Horizons in Ecotheology*, edited by Ernst M. Conradie et al., 31–50. London: Bloomsbury T. & T. Clark, 2014.

———. "Windows to the Divine Spirit: Between Species Encounters, Wild Justice and Image Bearing in Ecological Perspective." In *The Nature of Things: Rediscovering the Spiritual in God's Creation*, edited by Graham Buxton and Norman Habel, 157–69. Eugene, OR: Pickwick Publications, 2016.

———. "Wisdom, Justice and Environmental Decision-Making in a Biotechnological Age." *Ecotheology* 8 (2003) 173–92.

———. "The Wisdom of Fools? A Theo-Dramatic Interpretation of Deep Incarnation." In *Incarnation: On the Scope and Depth of Christology*, edited by Niels Gregersen, 177–202. Minneapolis: Fortress, 2015.

———. *The Wisdom of the Liminal: Evolution and Other Animals in Human Becoming*. Grand Rapids: Eerdmans, 2014.

Deane-Drummond, Celia, and David Clough. *Creaturely Theology: God, Humans and Other Animals*. London: SCM, 2009.

Deane-Drummond, Celia, and Heinrich Bedford-Strohm, eds. *Religion and Ecology in the Public Sphere*. London: T. & T. Clark/ Continuum, 2011.

Deane-Drummond, Celia, Rebecca Artinian Kaiser and David Clough, eds. *Animals as Religious Subjects: Transdisciplinary Perspectives*. T. & T. Clark Theology. London: Bloomsbury T. & T. Clark, 2013.

Deane-Drummond, Celia, Sigurd Bergmann and Markus Vogt, eds. *Religion in the Anthropocene*. Eugene, OR: Cascade Books, 2017.

DeWitt, Calvin B. "Creation's Environmental Challenge to Evangelical Christianity." In *The Care of Creation: Focusing Concern and Action*, edited by R. J. Berry, 60–73. Leicester, UK: Inter-Varsity, 2000.

The Earth Bible Team. "Guiding Ecojustice Principles." In *Readings from the Perspective of Earth*, edited by Norman Habel, 38–53. Earth Bible 1. Sheffield: Sheffield Academic, 2000.

Eaton, Heather. "Epilogue: A Spirituality of the Earth." In *The Nature of Things: Rediscovering the Spiritual in God's Creation*, edited by Graham Buxton and Norman Habel, 229–46. Eugene, OR: Pickwick Publications, 2016.

———. *Introducing Ecofeminist Theologies*. Introductions in Feminist Theology 12. London: Continuum, 2005.

———. *The Intellectual Journey of Thomas Berry: Imagining the Earth Community*. Lanham, MD: Lexington, 2014.

Edwards, Denis. *Ecology at the Heart of Faith*. Maryknoll, NY: Orbis, 2006.

Ehrman, Terrence. "Ecology: The Science of Connections." In *Everything Is Connected: Pope Francis' Ecological Vision in Laudato Si*, edited by Vincent Miller. London: Bloomsbury T. & T. Clark, 2017, *in press*.

Francis, Pope. *Laudato Si': On Care for Our Common Home. Encyclical Letter.* Huntington, IN: Our Sunday Visitor, 2015.

———. *Mass, Imposition of the Pallium and Bestowal of the Fisherman's Ring for the Beginning of the Petrine Ministry of the Bishop of Rome, Homily of Pope Francis*. (19 March 2013). http://www.vatican.va/holy_father/francesco/homilies/2013/documents/papa-francesco_20130319_omelia-inizio-pontificato_en.html.

Fuentes, Agustin. "Naturalcultural Encounters in Bali: Monkeys, Temples, Tourists and Ethnoprimatology." *Cultural Anthropology* 25 (2010) 600–624.

Goodin, Robert E. "Selling Environmental Indulgences." In *Environmental Ethics and Philosophy*, edited by John O'Neill et al., 493–515. Managing the Environment for Sustainable Development 6. Northampton, MA: Elgar 2001.

Gregersen, Niels Henrik. "Christology." In *Systematic Theology and Climate Change: Ecumenical Perspectives*, edited by Michael S. Northcott and Peter M. Scott, 33–50. New York: Routledge, 2014.

———. "The Cross of Christ in an Evolutionary World." *Dialog: A Journal of Theology* 40 (2001) 192–207.

———. *Incarnation: On the Scope and Depth of Christian Theology*. Minneapolis: Fortress, 2015.

Grey, Mary. *The Outrageous Pursuit of Hope: Prophetic Dreams for the Twenty First Century*. London: Darton, Longman & Todd, 2000.

———. *Sacred Longings: Ecofeminist Theology and Globalization*. London: SCM, 2003.

Griffin, Susan. *Women and Nature: The Roaring Inside Her*. London: Women's Press, 1984.

Grim, John, and Mary Evelyn Tucker. *Ecology and Religion*. Foundations of Contemporary Environmental Studies. Washington, DC: Island Press, 2014.

Gustafson, James M. *A Sense of the Divine: The Natural Environment from a Theocentric Perspective*. Edinburgh: T. & T. Clark, 1994.

Gutiérrez, Gustavo. *A Theology of Liberation: History, Politics, and Salvation*. Translated and edited by Sister Caridad Inda and John Eagleson. Rev. ed. Maryknoll, NY: Orbis, 1988.

Habel, Norman. "Where Can Wisdom Be Found? Re-Discovering Wisdom in God's Creation." In *The Nature of Things: Rediscovering the Spiritual in God's Creation*, edited by Graham Buxton and Norman Habel, 139–156. Eugene, OR: Pickwick Publications, 2016.

———. "Where Is the Voice of Earth in the Wisdom Literature?" In *The Earth Story in Wisdom Traditions*, edited by Norman Habel and Shirley Wurst, 23–34. The Earth Bible 3. Sheffield: Sheffield Academic, 2001.

Horrell, David. *The Bible and Environment: Towards a Critical Ecological Biblical Theology*. Durham, UK: Acumen, 2013.

Haught, John F. *The Promise of Nature: Ecology and Cosmic Purpose*. 1993. Reprint, Eugene, OR: Wipf & Stock, 2004.

Hutchinson, G. E. *The Ecological Theater and the Evolutionary Play*. New Haven: Yale University Press, 1965.

Intergovernmental Panel on Climate Change. "Summary for Policymakers." In *Climate Change 2014, Mitigation of Climate Change. Contribution of the Working Group III to the Fifth Assessment Report of the Intergovernmental Panel on Climate Change*, edited by O. Edenhofer et al. Cambridge: Cambridge University Press, 2014. http://mitigation2014.org/.

International Union for Conservation of Nature (IUCN). "New Bird Species and Giraffe Under Threat—IUCN Red List." December 8, 2016. https://www.iucn.org/news/new-bird-species-and-giraffe-under-threat-iucn-red-list.

Jackson, Wes. "The Agrarian Mind. Mere Nostalgia or Practical Necessity?" In *The Essential Agrarian Reader: The Future of Culture, Community and the Land*, edited by Norman Wirzba, 140–53. Berkeley: Counterpoint, 2004.

Jantzen, Grace. *Becoming Divine: Towards a Feminist Philosophy of Religion*. Manchester: Manchester University Press, 1998.

———. *God's World, God's Body*. Philadelphia: Westminster, 1984.

John Paul II, Pope. *Centesimus Annus*. London: Catholic Truth Society, 1991. http://www.vatican.va/holy_father/john_paul_ii/

encyclicals/documents/hf_jp-ii_enc_01051991_centesimus-annus_en.html/.

———. "General Audience." (26 January 2000). http://www.vatican.va/holy_father/john_paul_ii/audiences/2000/documents/hf_jp-ii_aud_20000126_en.html/.

John Paul II, Pope, and Patriarch Bartholomew I. *Common Declaration on Environmental Ethics.* (June 10, 2002). https://w2.vatican.va/content/john-paul-ii/en/speeches/2002/june/documents/hf_jp-ii_spe_20020610_venice-declaration.html.

Johnson, Elizabeth. "Jesus and the Cosmos: Soundings in Deep Christology." In *Incarnation: On the Scope and Depth of Christology*, edited by Niels Gregersen, 133–56. Minneapolis: Fortress, 2015.

Kim, Grace Ji-Sun, and Hilda P. Koster, eds. *Planetary Solidarity: Global Women's Voices on Christian Doctrine and Climate Justice.* Minneapolis: Fortress, 2017.

Locke, Piers. "Explorations in Ethnoelephantology: Social, Historical, and Ecological Intersections Between Asian Elephants and Humans." *Environment and Society: Advances in Research* 4 (2013) 79–97.

McCulloch, Gillian. *The Deconstruction of Dualism in Theology: With Special Reference to Ecofeminist Theology and New Age Spirituality.* Carlisle, UK: Paternoster, 2003.

McDonagh, Sean. *The Greening of the Church.* London: Chapman, 1990.

McFague, Sally. *The Body of God: An Ecological Theology.* Minneapolis: Fortress, 1993.

———. *Life Abundant: Rethinking Theology and Economy for a Planet in Peril.* Minneapolis: Fortress, 2001.

McKibben, Bill. *Oil and Honey: The Education of an Unlikely Activist.* New York: St. Martin's Griffin, 2014.

Minns, Denis. *Irenaeus: An Introduction.* London: Chapman, 1994.

Moltmann, Jürgen. *God in Creation: An Ecological Doctrine of Creation.* Translated by Margaret Kohl. Gifford Lectures 1984–1985. London: SCM, 1985.

Mora, Camilo, et al. "How Many Species Are There on Earth and in the Ocean?" *PLoS Biology* 9 (2011) 1–8.

Nash, James. "The Bible vs Biodiversity: The Case against Moral Argument from Scripture." *Journal for the Study of Religion, Nature and Culture* 3 (2009) 213–37.

Northcott, Michael S. *The Environment and Christian Ethics.* New Studies in Christian Ethics. Cambridge: Cambridge University Press, 1996.

———. *Place, Ecology and the Sacred: The Moral Geography of Sustainable Communities*. London: Bloomsbury Academic, 2015.
Nothwehr, Dawn M., ed. *Franciscan Theology of the Environment: An Introductory Reader* Quincy, IL: Franciscan, 2002.
Nussbaum, Martha. *Frontiers of Justice: Disability, Nationality, Species Membership*. Tanner Lectures on Human Values. Cambridge, MA: Belknap, 2006.
Painter, John. "Theology, Eschatology and the Prologue of John." *Scottish Journal of Theology* 46 (1993) 27–42.
Parsons, Susan. *The Ethics of Gender*. New Dimensions to Religious Ethics. Oxford: Blackwell, 2002.
Peppard, Christina Z. *Just Water: Theology, Ethics and the Global Water Crisis*. Maryknoll, NY: Orbis, 2014.
———. "Laudato Si' and Standing Rock: Water Justice and Indigenous Ecological Knowledge", in *Theology and Ecology across the Disciplines: On Care for Our Common Home*, edited by Celia Deane-Drummond and Rebecca Artinian Kaiser. London: Bloomsbury, 2018, *in press*.
Primavesi, Anne. *From Apocalypse to Genesis: Ecology, Feminism and Christianity*. Minneapolis: Fortress, 1991.
———. *Sacred Gaia: Holistic Theology and Earth Systems Science*. London: Routledge, 2000.
Rawls, John. *A Theory of Justice*. Cambridge: Harvard University Press, 1971.
Raworth, Kate. "A Safe and Just Space for Humanity: Can We Live within the Doughnut?" *Oxfam Policy and Practice: Climate Change and Resilience* 8 (2012) 1–26.
Rockström, Johan, et al. "A Safe Operating Space for Humanity." *Nature* 461 (2009) 472–75.
Rockström, Johan, et al. "Planetary Boundaries: Exploring a Safe Operating Space for Humanity." *Ecology and Society* 14 (2009) 32. http://www.ecologyandsociety.org/vol14/iss2/art32/.
Rolston, Holmes, III. *Genes, Genesis and God: Values and Their Origins in Natural and Human History*. Gifford Lectures 1997–1998. Cambridge: Cambridge University Press, 1999.
Ruether, Rosemary Radford. *Gaia and God: An Ecofeminist Theology of Earth Healing*. San Francisco: HarperSanFrancisco, 1992.
———. *Introducing Redemption in Christian Feminism*. Introductions in Feminist Theology 1. Sheffield: Sheffield Academic, 1998.
Rushton, Kathleen. "The Cosmology of John 1:1–14 and Its Implications for Ethical Action in this Ecological Age." *Colloquium* 45 (2013) 137–53.

Sakimoto, Philip J. "Understanding the Science of Climate Change." In *Theology and Ecology across the Disciplines: On Care for Our Common Home*, edited by Celia Deane-Drummond and Rebecca Artinian Kaiser. London: Bloomsbury, 2018, *in press*.

Schama, Simon. *Landscape and Memory*. London: Fontana, 1996.

Schaeffer, Jame. *Theological Foundations for Environmental Ethics: Reconstructing Patristic and Medieval Concepts*. Washington, DC: Georgetown University Press, 2009.

Scott, Peter. *A Political Theology of Nature*. Cambridge Studies in Christian Doctrine 9. Cambridge: Cambridge University Press, 2003.

Sen, Amartya. *The Idea of Justice*. London: Allen Lane/Penguin, 2009.

Sideris, Lisa. *Environmental Ethics, Ecological Theology and Natural Selection*. Columbia Series in Science and Religion. New York: Columbia University Press, 2003.

Smith, Graham. *Deliberative Democracy and the Environment*. Environmental Politics. London: Routledge 2003.

Steffen, Will, et al. "Planetary Boundaries: Guiding Human Development on a Changing Planet." *Science* 347 (2015). DOI: 10.1126/science.1259855

Taylor, Bron, and Michael Zimmermann. "Deep Ecology." In *Encyclopedia of Religion and Nature*, edited by Bron Taylor, 1:456–60. London: Continuum, 2005.

Tucker, Mary Evelyn, and Brian Swimme. *Journey of the Universe*. New Haven: Yale University Press, 2011.

U.S. Environmental Protection Agency. "Climate Change." August 9, 2016. http://www3.epa.gov/climatechange/ghgemissions.gases.html.

Walls, Laura. "Cosmos." In *Keywords for Environmental Studies*, edited by Joni Adamson et al., 47–49. New York: New York University Press, 2016.

White, Lynn. "The Historical Roots of Our Ecologic Crisis." *Science* 155 (1967) 1203–7.

White, Sarah, and Romy Tiongco. *Doing Theology and Development: Meeting the Challenge of Poverty*. Edinburgh: St. Andrew's, 1997.

Wilson, Edward O. "A Cubic Foot." *National Geographic* (February 2010).http://ngm.nationalgeographic.com/2010/02/cubic-foot/wilson-text/1.

Wirzba, Norman. *Living the Sabbath: Discovering the Rhythms of Rest and Delight*. Grand Rapids: Brazos, 2006.

———. *The Paradise of God: Renewing Religion in an Ecological Age*. Oxford: Oxford University Press, 2003.

GENERAL INDEX

General Index

SCRIPTURE INDEX

Romans

Philippians

Colossians

1 Timothy

Hebrews

Revelation